THE INTERNET AND SOCIETY

A Reference Handbook

Other Titles in ABC-CLIO's
CONTEMPORARY
WORLD ISSUES
Series

Books in the Contemporary World Issues series address vital issues in today's society such as genetic engineering, pollution, and biodiversity. Written by professional writers, scholars, and nonacademic experts, these books are authoritative, clearly written, up-to-date, and objective. They provide a good starting point for research by high school and college students, scholars, and general readers as well as by legislators, business people, activists, and others.

Each book, carefully organized and easy to use, contains an overview of the subject, a detailed chronology, biographical sketches, facts and data and/or documents and other primary-source material, a directory of organizations and agencies, annotated lists of print and nonprint resources, and an index.

Readers of books in the Contemporary World Issues series will find the information they need in order to have a better understanding of the social, political, environmental, and economic issues facing the world today.

THE INTERNET AND SOCIETY

A Reference Handbook

Bernadette H. Schell

CONTEMPORARY WORLD ISSUES

A B C · C L I O

Santa Barbara, California
Denver, Colorado
Oxford, England

Library of Congress Cataloging-in-Publication Data

Schell, Bernadette H. (Bernadette Hlubik), 1952–
    The Internet and society : a reference handbook / Bernadette H. Schell.
        p.   cm. — (ABC-CLIO's contemporary world issues)
    Includes bibliographical references and index.
    ISBN-10: 1-59884-031-2 (hardcover : alk. paper)
    ISBN-10: 1-59884-032-0 (ebook)
    ISBN-13: 978-1-59884-031-5 (hardcover : alk. paper)
    ISBN-13: 978-1-59884-032-2 (ebook)
    1. Internet—Social aspects.   2. Internet—History.   3. Internet—Biography.   4. Internet—Law and legislation.   5. Computer crimes. 6. Digital divide.   I. Title.
HM851.S25 2007
303.48'33—dc22
                                                                2006036122

11   10   09   08   07   1 2 3 4 5 6 7 8 10

This book is also available on the World Wide Web as an eBook. Visit abc-clio.com for details.

ABC-CLIO, Inc.
130 Cremona Drive, P.O. Box 1911
Santa Barbara, California 93116–1911

This book is printed on acid-free paper. ∞
Manufactured in the United States of America.

Contents

Tables

Preface

As part of the Contemporary World Issues series, *Internet and Society: A Reference Handbook* is intended for a general audience with no technical knowledge of networking. Its purpose is to help readers better understand the controversial political, social, technical, legal, and economic issues surrounding the Internet—or the Net, as it is commonly called. Because of its nontechnical writing style, this book should be useful to college and university students in any discipline, not only those in sociology or computer science. Also, because this book is consistent in form with the *Contemporary World Issues* series, it is relevant to a broader adult audience, including professionals and lawyers wanting to know more about how the Internet operates and its impact on society.

Since its appearance in the United States in the second half of the twentieth century, the Internet has been the subject of arduous study from different disciplines, including the social sciences, business, law, computer science, and political science. In the last few decades, when the Internet started to expand at unprecedented rates and with different socioeconomic interests involved, the Net's impact on global citizens' daily lives has been profound. Representing a solid source of economic, social, and political information, the Internet has, without question, become one of the most important means of international communications (including online activism—known as hacktivism) and for advancing economies around the world. Moreover, the complex infrastructure of the World Wide Web has facilitated a number of common activities and, thus, has provided a more level political and economic "playing field" for citizens in both developed and developing countries.

With the advent of the Internet in the 1960s, a variety of positive services have been made available to citizens around the globe, ranging from communications—such as instant messaging and telephony—to rapid, real-time online transactions—such as e-commerce, Internet banking, online gaming, political activism, and online voting. Through the use of handheld Internet wireless devices, physicians have readily been able to access patients' health histories and diagnostic records without having to rely on time-delaying courier services; young billionaires have been created with the development of "Google-like" search engines; and governments around the globe have used the Internet to collect homeland security intelligence as a means of keeping their citizens safe.

However, with the growth of and diversity in Internet traffic, a dark side to the Internet has surfaced in recent decades, particularly since the late 1980s. That growth has played an important role in advancing abuses of available online services. Thus, tech-savvy criminals have increasingly made use of the Internet to perpetrate online crimes—which has led to a marked worldwide rise in cases of child pornography, identity theft, intellectual property theft, worm and virus infestations of business and home computers (resulting in a drastic reduction in computers' information-processing effectiveness), and online fraud involving e-commerce, voting, and gaming. Also, cell phone—owning citizens—including famous ones such as Paris Hilton—have discovered their stored personal messages and photos displayed over the Internet for millions of eyes to peruse. According to recent estimates, the cost to victims of malicious computer attacks (known in the security field as "cracks") have totaled more than $10 billion since the turn of the millennium, including cellular phone exploits (IBM Research, 2006).

Because of illegal access by cyber-criminals to online data files, citizens have become increasingly fearful that their personal health histories, their banking transactions transmitted over the Net, and their online voting results are vulnerable to destruction or alteration. Also, business executives have become increasingly concerned, not only that their computer networks will be cracked by tech-savvy outsiders but also that insider employees will either destroy critical business information when they leave the corporation—or they will sell industrial secrets and proprietary information to competitors over the Net.

As a result of the concerns and expectations surrounding on-line privacy, security, and trust issues of citizens, government officials, financial institution executives, and businesses leaders, various technical and legal solutions have emerged in the United States and elsewhere to combat Internet abuses. These laws—often controversial themselves—have included the USA Patriot Act, the Digital Millennium Copyright Act (DMCA), the Electronic Communications Privacy Act, the Foreign Intelligence Surveillance Act, the Prosecutorial Remedies and Tools against the Exploitation of Children Today (or PROTECT) Act, the Health Insurance Portability and Accountability Act (HIPPA), and the Gramm-Leach Bliley Act.

Besides all of these pluses and minuses, the Internet has had a tremendous economic and political impact on underdeveloped and developing countries, particularly in China, India, Russia, and Pakistan. Compared with the development of the Net in North America—where companies like Yahoo, Inc., have been able to succeed because they got onto the Information Highway early on, had plenty of financing to hire the very best professionals, and had a supportive government—in China, as a case in point, Baidu, the Chinese Internet search engine that emerged as a growing success story in 2005, has been five years in the making, and has often been at odds with the government. Although relatively unknown outside of China until now, Baidu was founded by two Chinese gurus who had previously worked for U.S. technology firms. The entrepreneurs who own the company claim that they have a strong following of more than 100 million Web surfers in China, but they openly admit that, relative to the open surfing privileges given North American Web surfers, the Chinese government censors whatever content they feel would corrupt the morals of its citizens (In Brief, 2005; Vage, 2005).

It is interesting to note, however, that the Internet did not develop from "thin air." Rather, its development and growth over a remarkably short time occurred as a result of some very talented Americans and Canadians. Any list of Internet notables includes some of the world's wealthiest innovators, such as U.S.-born Bill Gates, who in 1975 started Microsoft. A Canadian innovator who is at the top of the list of notables is Tim Bray, who helped create an index called Extensible Markup Language (or XML)—thus allowing computer programmers to attach universal codes (or tags) to distinguish, say, a business name from a phone number and thereby aid in the boom for electronic commerce (that is, business conducted over the Net). Also, Canada's Stephen Cook, a University of Toronto professor, helped to develop the field of cryptography—the underlying basis for relatively safe online Web purchases. Other notables include Mers Kutt, who in 1973 started Micro Computer Machines—thereby introducing the world's first personal computer. Then there is James Gosling, credited with producing a Web revolution allowing the World Wide Web (typically abbreviated "WWW") to move from a static environment to an interactive, user-friendly one in which online surfers around the world can post messages, order books in a matter of minutes without leaving home, and download songs from the comfort of their bedrooms. Other notable North Americans include Dennis M. Ritchie of UNIX operating system fame; Vinton Cerf, involved with Internet protocols; Shawn Fanning, known for the controversial online song-swapping Napster; Rail Tomlinson, an e-mail contributor; and Tim Berners-Lee, a World Wide Web pioneer.

On the darker side—at least for parts of their younger lives and to name just two of many—are North American cyber-criminals Kevin Mitnick—an American who spent years in prison for online Denial of Service (DoS) exploits but who is today considered one of the world's finest Internet security professionals—and Mafiaboy, a young Canadian who, unlike Mitnick, escaped prison for committing similar online exploits but who has yet to make it as a successful online security guru.

With the growth of the Internet came the monitoring of the critical workings and safety of this massive network by organizations and agencies such as the American Registry for Internet Numbers, the Asia Pacific Network Information Center, the Central Intelligence Agency, the CERT Coordination Center, the Cyber Warning and Information Network, the CyberAngels, the Defense Intelligence Agency, the Department of Homeland Security, the Department of Justice in the United States, the Electronic Frontier Foundation (EFF), the Federal Bureau of Investigation (FBI), the Federal Trade Commission (FTC), the Internet Assigned Numbers Authority, the Internet Corporation for Assigned Names and Numbers, the Internet Fraud Complaint Center—to name just a few.

This book details the impact that the Internet has had on society—the good and the not so good—from its beginnings to the present. Readers will be better able to understand how to optimize the opportunities that the Internet provides as well as how to best protect their property and their persons from harm when riding on this Information Highway. The chapters are as follows:

Chapter 1, "Background and History," defines what the Internet is, outlines the history of the Internet in the United States and elsewhere, and discusses the Net's impact socially, politically, technically, and economically. A global perspective is taken, and the positive impact is emphasized—especially early on. Some important evolutionary changes of the Internet described include the ARPAnet (1960s), electronic mail (1971), remote login (1972), file transfer (from the basic file transfer applications of the 1970s to the sophisticated file-sharing applications of our day), the emergence of the World Wide Web, the Universal Resource Locator (1989–1991), the birth of Google and of Google's billionaires post-2000, and the future of Voice over Internet Protocol (VoIP). Following positive details such as these, a chronology of key events related to the growth of the Internet is presented at the end of the chapter.

Chapter 2, "Problems, Controversies, and Solutions," discusses the controversies surrounding the Internet and possible solutions. With the growth of and diversity in Internet traffic, a dark side to the Internet has surfaced, particularly since the late 1980s. The first half of Chapter 2 explores a number of privacy, security, and trust issues of concern to various segments of society, techies and nontechies. A number of crimes perpetrated by abusing the Internet infrastructure are described, such as illegal file transfer (for example, copyrighted material abuse, intellectual property abuse), transfer of censured material (such as child pornography), identity theft, and fraud (for example, through pharming and online auction abuses). Similar to e-mail phishing, pharming tries to get personal or private and typically money-related information from online users through domain spoofing. So, rather than being spammed with e-mail requests prompting online users to visit various Websites appearing to be legitimate, pharming poisons a DNS server by putting in fake information to the DNS server. The result is that the online user's request is redirected to some other Website other than that requested. The second half of Chapter 2 explains various laws passed in the United States and elsewhere to reduce or prevent common Internet crimes, as well as some technical solutions.

Chapter 3, "Worldwide Perspective," discusses the economic and political impact of the Internet on countries outside of North

America, particularly in the underdeveloped and developing nations and including China, India, Russia, and Pakistan.

Chapter 4, "Chronology," details interesting events from the foundational start of the Internet to the present.

Chapter 5, "Biographical Sketches," includes twenty biographical sketches of people who have significantly contributed to the development of the Internet, or who have used the Internet to execute harm on people or property.

Chapter 6, "Facts and Data," examines the laws and legal cases in the United States and elsewhere relating to Internet abuses.

Chapter 7, "Directory of Organizations," describes organizations created in the United States and elsewhere for setting Internet-related policies and for curbing Internet-related abuses.

Chapter 8, "Print and Nonprint Resources," provides additional print and nonprint resources related to the Internet.

The glossary provides Internet terminology to assist readers in their understanding of the basic workings of the Internet.

Resources

IBM Research. "Global Security Analysis Lab: Fact Sheet." IBM Research, http://domino.research.ibm.com/comm/pr.nsf/pages/rsc.gsal.html? Open&printable (accessed January 16, 2006).

In Brief. "China Cracks Down on Public Internet." *Globe and Mail*, February 17, 2005, p. B10.

Vage, Lars. "China's Search Engine Censorship Continues." Internet-Brus, http://www.pandia.com/sw–2005/09-china.html (accessed January 27, 2005).

1

Background and History

This chapter defines what the Internet is, outlines the history of the Internet in the United States—from its beginnings in the 1960s to its rapid growth globally in the 1990s—and discusses the Internet's impact socially, politically, technically, and economically over four decades. The positive effects are emphasized near the beginning, but the negative side is introduced later on, because that is the reality associated with Internet development. A section on how the Internet works follows. The chapter closes with a look at future developments of the Internet. A chronology of key events related to the growth of the Internet is presented at the end of the chapter.

What the Internet and an Intranet Are— and the Internet's Colorful History

What the Internet Is

An "internet," generally speaking, is a network connecting computer systems. The Internet, as we know it today, is also called the World Wide Web—which had its seeds planted in the United States back in 1969. At that time, "the Internet" was a designated high-speed network built by the U.S. Department of Defense as a digital communications experiment. By linking hundreds of universities, defense contractors, and research labs, ARPAnet—the computer network belonging to the U.S. Advanced Research Projects Agency (ARPA)—let Artificial Intelligence (AI) researchers in dispersed areas exchange critical information with incredible speed and flexibility. Simply put, AI is the branch of computer science concerned with making computers behave like humans by modeling on computers human thoughts. Sometimes, AI is meant to solve a problem that a person can solve but do it more efficiently using a computer.

Although this ARPA "Internet" was reserved exclusively for the U.S. military and a very select group of contractors and universities involved in defense-related research, its capability advanced the field of Information Technology (IT). Instead of working in isolated pockets, key agents involved in keeping the United States safe were now able to communicate through the electronic highway as networked tribes rather than having to fly from one destination to another to complete the information exchange. It is interesting to note that this networked tribe phenomenon still exists in the Computer Underground (CU), or hacking world, today.

What an Intranet Is

An Intranet, or, more properly, Intranet site, is an information system internal to an organization built with Web-based technology. An Intranet site is often called a "portal," and it typically exists in large companies having in excess of 15,000 employees. It is considered by many to be an IT luxury.

An Intranet site is actually a mini-Internet accessed through a Web browser and is typically run on private Local Area Networks (LANs) rather than on public Web servers like the World Wide Web. A Web browser interprets HTML (Hyper Text Markup Language), the programming language used to code Web pages on the Internet, into words and graphics so that users can view the pages in their intended layout and rendering.

Intranet sites have a number of functions, but most are used to keep employees informed about a company's upcoming events, to distribute software or company newsletters online, and to provide routine company information online. New Intranet site software made by companies such as the Microsoft Corporation and Plumtree Software, Inc., has made the technology affordable in 2006 even for small and medium-size companies (Palmer, 2005).

The Colorful History of the Internet and Its Increasingly Problematic Future

A number of factors have pushed the development of the powerful Internet that society enjoys today—businesses, universities, governments, medical institutions, financial institutions, adults, and children. No single factor produced the World Wide Web that today connects so many global networks. Groundbreaking events in hardware (the mechanical parts composing the computer network) and software (the programming language running the network and doing the necessary computations) creation have taken place, and these developments have had huge social and economic impacts on society.

The 1940s, 1950s, and 1960s.   In the late 1940s and 1950s, when computers were made with 10,000 vacuum tubes and occupied more than 93 square meters of space, there was indeed a limit to how big the cyber beasts could be, for they could overheat and explode. There were other problems with those early computers. The vacuum tubes could leak; the metal that emitted electrons in the vacuum tubes burned out; and all of those tubes required tons of power to run. Major improvements came in computer hardware technology with the development of transistors, invented by John Bardeen and Walter Brattain in 1947 and 1948 at Bell Telephone Laboratories, and the development of integrated circuits, invented by Jack Kilby at Texas Instruments and Robert Noyce at Fairchild Camera in 1958 and 1959. Integrated circuits resolved a number of vacuum-tube technology's problems and did much to further the the development of smaller computers with greater power. With this improved technology, instead of making transistors one by one, several transistors could be made at the same time and on the same piece of semiconductor. In addition to transistors, other electric components such as resistors, capacitors, and diodes could be made using the same process and materials (Haviland, 2005).

It is important to note that since the 1960s, the number of transistors per unit area has been doubling every one-and-a-half years—thus increasing computing power. This amazing progression of circuit fabrication is called Moore's Law, named after Gordon Moore, a pioneer in the integrated circuit field and founder of the Intel Corporation (ibid.).

Indeed, the art of computer networking dates back to the 1960s. As computer use was on the increase back then, the question of how to hook up and share the data between computers remained. To this end, the first revolutionary packet-switching technique was invented by Leonard Kleinrock. The idea soon gained in popularity, and gifted people like Paul Baran at the Rand Institute began investigating how to use packet switching for secure voice over military networks (Edge, 2002).

One of the early critical events making networking on a broader scale possible occurred in 1969, when Bell Laboratory IT gurus Dennis Ritchie and Ken Thompson developed the computer software standard operating system UNIX. At that time, UNIX was (and still is) considered a thing of beauty, because its standard user and programming interfaces assisted with computing, word processing, and networking. It must be emphasized, however, that back in the 1960s, computer hardware was huge and was stored in air-conditioned areas to keep it from overheating.

The 1970s. In 1970, only an estimated 100,000 computers were in use in the United States; few of them were networked, and most of them were found in businesses, in government offices, and in universities. Computer users keypunched data cards and carried them in an orderly fashion in boxes, then dumped the cards into the computer for processing.

In 1971, Ray Tomlinson gave society one of the greatest communications gifts in Internet history: electronic mail, or e-mail. Tomlinson, a graduate of the Massachusetts Institute of Technology (MIT), is one of the forefathers of the Internet, for he worked on ARPAnet. He is also the man who chose the @ sign for e-mail, still in use today to mean "at," as in Bernadette Schell@uoit.ca (Gaudin, 2002).

In 1971, Stephen Cook, a Canadian professor at the University of Toronto, published Cook's Theorem, identifying a large class of computational search problems that at the time would have taken even the largest, highest-power computers millions of years to solve. This theorem helped lead to the development of the field of cryptography—critical now for secure online business transactions. Today, after an Internet purchaser places an online order, the retailer selling the merchandise has a computer system that encrypts, or encodes, the data using a complex algorithm. Back in the 1970s, Cook's Theorem predicted that a cyber-criminal would need to spend a lifetime finding the solution for breaking encoded data. This theorem alone has saved billions of dollars in online transaction thefts. For example, in 2004, 16 million fellow Canadians completed online transactions through the Net, and 12 million of them completed their banking transactions through the Net (Brearton, 2005).

In 1972, the National Center for Supercomputing Applications (NCSA) created the telnet application for remote login, providing an easier means for users to log into a remote machine. Also in 1972, David Boggs and Robert Metcalfe invented Ethernet at the Xerox Corporation in California. Ethernet back then used a passive bus operated at 10 megabits per second (Mbps) and a 50-ohm coaxial cable to connect computers in the Local Area Network (LAN). It was not until 1985 that the U.S. Institute of Electrical and Electronic Engineers (IEEE) developed standards for LANs—known as the IEEE 802 standards, which today form the basis of most networks around the world. Although a single LAN can have as many as 1,024 attached computer systems, in practice most LANs have far fewer.

In 1973 the File Transfer Protocol (FTP) was developed, simplifying the transfer of information between networked machines. Also in the 1970s, Dennis Ritchie invented a new programming language called "C," which, like UNIX in the operating system world, was designed to be pleasant, unconstraining, and flexible. By the end of the decade, the whole environment had successfully been ported to machines of different types.

In the early 1970s, computer hardware specialists were "hung up" on the idea that bigger meant more powerful—at least as far as computers went. However, Canadian mathematician and businessman Mers Kutt had a novel idea; he thought that smaller is better! In 1973, this pioneer founded Toronto, Ontario-based Micro Computer Machines, which released onto the market the world's first Personal Computer (PC). This PC looked a lot more like a typewriter than a laptop; it weighed 9 kilograms, and it had tape drives to store information. It could hold only 8 kilobytes of memory—compared with today's PCs, which can hold about 30,000 times that amount—and the display console could hold only thirty-six or so characters. It cost around $5,000 and got its power from a small chip rather than vacuum tubes. Today, the company is selling millions of highly improved, vastly more powerful Personal Computers (ibid.).

In 1975, Apple Computer, Inc., was founded by two members of California's Homebrew Computer Club: Steve Jobs and Steve Wozniak. Once the Apple Computer and the simplistic and user-friendly BASIC computer language were purchased by tech-savvy types, home use of computers—which is common in developed countries today—was on its way to becoming a reality.

Also in 1975, William Henry Gates, III—Bill Gates, as the world knows him—and Paul Allen founded the Microsoft Corporation. Nowadays, *Microsoft* is a word that even children understand and "crackers" (people who intentionally break into computer systems to cause harm, to take revenge, or for personal financial gain) often detest—a testament to the software company's success.

Secure transactions using the Internet had their roots planted in 1976, when Whitfield Diffie and Martin Hellman created the Diffie-Hellman Public-Key Algorithm, or DH. The DH, an algorithm used in many secure protocols on the Internet, is now celebrating thirty years of use. An algorithm is a set of rules and procedures for resolving a mathematical or logical problem, much as a recipe in a cookbook helps baffled cooks to solve a meal-preparation issue. Actually, a computer program can be thought of as an elaborate algorithm. In computer science, an algorithm—the term is believed to stem from the name of a mathematician at the Royal Court in Baghdad in the ninth century—is a mathematical procedure for solving a recurrent problem. Often inventors of algorithms seek patents to ensure the safety and proper recognition of their Intellectual Property (IP). Information security professionals, in particular, work with cryptographic algorithms to encrypt, or encode, messages sent over the Internet.

By the end of the 1970s, the one positive thing missing from the cyber community that could serve a much-needed social function was a form of networking social club. In 1978, the void was filled by two men from Chicago, Randy Suess and Ward Christiensen. This pair created the first computer Bulletin Board System (BBS), the forerunner of today's Internet chat rooms.

It was also at the end of the 1970s that the Transmission Control Protocol (TCP) was split into TCP and IP (Internet Protocol). Without getting too technical, it is important to know that information is sent over the Internet in information packets that are disassembled and then reassembled. The TCP and IP are critical for the proper transmission and routing of these information packets. Widely known as one of the Fathers of the Internet, Vinton Cerf is the codesigner of the TCP/IP protocols and the architecture of the Internet. In December 1997, President Clinton presented the U.S. National Medal of Technology to Vinton Cerf and Robert Kahn, for helping to develop the Internet (ICANN, 2005).

The 1980s. In 1981, Personal Computers (PCs) finally became affordable. IBM (International Business Machines) announced their new model, stand-alone computer, which was actually called "the PC." Also, the Commodore 64 computer (known to many techies as the "Commie 64") was released, as was the TRS-80 computer (known to many as the "Trash-S"). It is interesting to note that in 1981, fewer than 300 computers were linked to the Internet.

In 1982, a group of talented UNIX hackers from Stanford University and the University of California at Berkeley—Scott McNealy, Bill Joy, Andy Bechtolsheim, and Vinod Khosla—founded Sun Microsystems Incorporated on the foundation that UNIX running on relatively cheap hardware would prove to be a perfect solution for a wide range of applications. These visionaries seemed to know what they were doing, for their Sun Microsystem Networks increasingly replaced older, less efficient computer systems like the VAX and other time-sharing systems that consumed much space in corporations and universities across North America.

On January 13, 2006, these four Sun Microsystem visionaries were treated like famous rock stars when they appeared at the Computer History Museum in Mountain View, California. Their message was this: Sun Microsystems Incorporated has remained loyal to its original philosophy of sharing technology and enabling innovation. McNealy, the company's chair and CEO, credited Bill Joy—the UNIX guru—with the concept of online community development, resulting from work at the University of Berkeley related to Berkeley Software Distribution. Over the years, McNealy noted, Sun Microsystems has built a huge cyber community around Java technology, in particular. McNealy further noted that the company's Java Virtual Machine, a free tool used to run Java applications on Microsoft Windows software (which is not free), is currently downloaded 20 million times each month. McNealy also affirmed that there are more than 3 billion Java technology–enabled devices currently available to techies and nontechies, including DVD players and cellular phones. This fact by itself shows how innovative products developed by innovative geniuses have helped to transform the world.

Bechtolsheim, now senior vice president of the company's Network Systems Group, added that Sun's growth is far from over. He maintained that there are many more business opportunities around Sun's open source and free products for the industry as a whole than there are with Microsoft's costly technologies (Sun Microsystems, Inc., 2006). [A Webcast of their talk that focused on how critical themes from their twenty-four-year history illuminate technology's very positive future is available at http://www.sun.com/2006–0113/feature/index.html.]

Other creative people did artistic things on the computer back in 1982. It was in this year that Scott Fahlman typed the first online smiley, known and used by adults and children today in their e-mail transmissions and created with the three simple computer board key presses :-). It was also then that William Gibson coined the term *cyberspace*. On a more technical note, it was in 1982 that the Simple Mail Transfer Protocol (SMTP)—which relates to how e-mail is transmitted between host computers and users—was published. The following year, the Telnet protocol was published, allowing technically inclined individuals familiar with UNIX operating system software to log onto other computers on the Internet.

In 1984, other interesting creative events occurred in the United States. For one, *2600: The Hacker Quarterly* magazine emerged—a particularly exciting medium for computer hackers interested in technical and political issues surrounding the computer world. This magazine—and today's Website of the same name founded by Eric Corley (a.k.a. Emmanuel Goldstein)—are still hugely popular with those in the Computer Underground. It was also in 1984 that Steven Levy's book *Hackers: Heroes of the Computer Revolution* was released, detailing the Hacker Ethic, a source still guiding neophyte hackers to this day. Also in 1984, Fred Cohen introduced the term *computer virus*, a cyberspace vandal that can cause corruption of and possible erasing of information stored in computer files.

In 1984, Cisco Systems, Inc., was begun by a small number of computer scientists at Stanford University, and to this day the company remains committed to developing Internet Protocol–based networking technologies, particularly in the areas of routing and switches. It was also about this time that the American Richard Stallman began constructing a clone of UNIX, written in the "C" programming language and available to those in the "wired world" for free. His project, called GNU (meaning Gnu's Not Unix) operating system, is popular among computer hackers today.

In 1984, there were also creative advances in knowledge on the Canadian front. At the University of Montreal, Gilles Brassard and fellow researcher Charles Bennett released an academic paper detailing how quantum physics could be used to create unbreakable codes using quantum cryptography. Without getting too technical, this form of data encoding uses subatomic particles transmitted as individual pulses of light as the key for unlocking succeeding messages. The estimated return on this one piece of Intellectual Property used in commercial applications today is quoted to be about U.S. $15 billion (Brearton, 2005).

In 1985, the online hacker "zine" known as *Phrack* was first published by Craig Neidor (a.k.a. Knight Lightning) and Randy Tischler (a.k.a. Taran King) on the *Phrack* Website. For twenty years, the online magazine had some enlightening pieces on topics such as anarchy, cryptography, reverse-engineering, and numerous other features of high-tech interest. Then, on January 24, 2005, *Phrack* announced that number 63 would be their final release. The magazine was distributed at the DefCon hacking gathering in Las Vegas in July (phrackstaff@phrack.org, 2005).

It was also in 1985 that the first Website domain name was registered and assigned, known as www.Symbolics.com, a privately held company that acquired the Intellectual Property of the previously publicly held company known as Symbolics, Inc. The company produces special-purpose computer systems for running state-of-the-art object-oriented software programs written in the Lisp computer language (Reti, 2006).

It was in 1985 that America Online (AOL)—a popular online search engine today—was incorporated under its original name of Quantum Computer Services, and in 1985, the Free Software Foundation (FSF) was founded by Richard Stallman. FSF then and now has been committed to giving computer users permission to use, study, copy, change, and redistribute computer programs. It is important to note that over the past twenty years, the FSF has promoted the development of free software and has enhanced awareness in the computer community and more broadly in society about the ethical and political issues associated with using free software.

It was also in 1985 that the word *hacker* was introduced into the mainstream in the United States. The word was coined by good folks in the Computer Underground—known as the White Hats—who developed software and hardware, tested them for flaws, and then found fixes (known as patches). There was never any negative connotation attached to the word; it meant simply a person who enjoyed learning the details of computer systems and how to stretch their capabilities. However, the news media kept using the word *hacker* incorrectly, to indicate someone who broke into computer systems and caused property harm—often for personal financial gain or revenge. In the Computer Underground, people with malicious intent are known as "crackers" or "Black Hats," but the press repeatedly failed to make the correction. This incorrect usage of the word by the North American and the international press continues to this day, despite the White Hats's continued efforts. Interestingly, the original use of the word *hacker* had nothing at all to do with computers. *Hacker* is a Yiddish word that means "inept furniture maker."

In 1986, the term "criminal hacker" was first alluded to in the British press. It was triggered with the convictions of Robert Schifreen and Steven Gold in April of that year. The pair broke into a text information retrieval system operated by BT Prestel and left a greeting for his Royal Highness the Duke of Edinburgh on his BT Prestel network mailbox. The pair were convicted on a number of criminal charges under the British Forgery and Counterfeiting Act of 1981. The case led to the eventual passage in Britain of the Computer Misuse Act of 1990, meant to prevent unauthorized access to computer systems and to stop cybercriminals from impairing or hindering access to data stored on a computer or system.

It was also in 1986 that the U.S. Congress passed the Computer Fraud and Abuse Act, meant to counteract fraud and associated activity aimed at or completed with computers. The act was amended in 1994, 1996, and 2001 by the USA Patriot Act—itself a very controversial piece of legislation for reasons that will be discussed later in this book. Also in 1986, the Internet Engineering Task Force (IETF) was formed to act as a technical coordination forum for those who worked on ARPAnet, on the U.S. Defense Data Network (DDN), and on the Internet core gateway system.

In 1988, the public first realized the damage Internet worms could cause when Robert Morris, Jr., accidentally unleashed one that he had developed while studying at Cornell University. Later called "the Morris worm," it infected and subsequently crashed thousands of networked computers. The graduate student was not sentenced to prison for this exploit but was sentenced to three years on probation and told to complete 400 hours of community service. Morris, a university professor today, was also ordered to pay a hefty $10,500 fine.

The same year, and largely as a reaction to the Morris worm incident, the Computer Emergency Response Team (CERT)/CERT Coordination Center was created in the United States. Located at Carnegie Mellon University, the center's function was to coordinate communications among homeland security experts during security emergencies.

It was also in 1988 that Kevin Mitnick—a famous, repeat cracking offender who was imprisoned a number of times and has now become a computer security expert and author—secretly monitored the e-mail of computer security officials at MCI (a leading U.S. communications provider company) and DEC (Digital Equipment Corporation, a pioneering company in the U.S. computer industry). For these exploits, he was convicted of damaging computers and stealing software. Mitnick was sentenced to one year in prison.

Also in 1988, the U.S. Secret Service began their investigations of those in the Computer Underground. The Secret Service secretly videotaped the SummerCon hacking convention attendees, suspecting that not all hacker activities were "White Hat" in nature.

In 1989, Herbert Zinn (a.k.a. Shadowhawk) was the first minor to be convicted for violating the U.S. Computer Fraud and Abuse Act of 1986. Zinn cracked the American Telephone and Telegraph (AT&T) computer network and the U.S. Department of Defense network, destroying files estimated to be worth $174,000. He copied electronic programs allegedly worth millions of dollars and published online passwords and instructions for exploiting computer systems. At the age of sixteen, Zinn was imprisoned for nine months and fined $10,000.

Also in 1989, a group of West German computer crackers led by Karl Koch (a member of the Chaos Computer Club) made international news headlines—labeled as the first cyber espionage case—after they broke into the U.S. government's computer network and then sold the operating system source code to the Soviet KGB, the agency in charge of Soviet state security.

By 1989 only about 90,000 computers were linked to the Internet. By that time, however, techies were excited about URLs. A URL, or Uniform Resource Locator, is the complete location of an Internet item a user may want to receive. It is an absolute reference, and it is synonymous with the Internet terms *location* and *Internet address*. Each Web page has a unique URL. Online users may instantly view a Web page if they know the URL. As a case in point, the URL for the University of Ontario is http://www.uoit.ca. The term "Http://" defines the Internet protocol being used. Other protocols are "gopher://" and "ftp://." The host's name—www.uoit.ca—is the location the user is interested in. What then follows is more detailed information, such as a directory or folder—a domain name typically enclosed by "/". An example would be http://www.uoit.ca/business/lecture1.htm. Of particular interest would be the business directory or folder and the lecture 1 individual file.

The 1990s. The year 1990 is a very significant one in terms of Internet development on a worldwide scale. It was then that Tim Berners-Lee and Robert Cailliau developed the protocols that became the World Wide Web (www) while working at the Cern laboratory in Geneva, Switzerland.

Tim Berners-Lee's present-day Website gives some interesting insights into this remarkable development. He notes that the original Web browser, or more appropriately browser-editor, was called "WorldWideWeb" (because, he says, that's the way it looked back then—very one dimensional). Later it was renamed "Nexus" to prevent confusion between the software program and the more abstract information space—currently spelled "World Wide Web" (with spaces). The reason, says Berners-Lee, that he chose the word "Nexus" is that he wrote the protocol using a NeXT computer. By using this protocol on his NeXT computer, Berners-Lee admits that he was able to accomplish work in a couple of months that would have taken him a year utilizing other platforms (Berners-Lee, 2006).

It is important to emphasize that until 1991, the Internet was restricted in the United States to linking the military and restricted university computers. In 1991 the ban preventing Internet access for businesses was lifted, thus making possible business transactions online.

In the early 1990s, Internet growth on a homeowner scale can be attributed to the fact that many more tech-savvy individuals could afford to buy PCs similar in power and storage capacity to the clunky computer systems of the 1980s. The reason for growth on this plane was the availability of newer, lower-cost, higher-performing PCs having Intel 386 chips. Despite this good news, affordable software was still a problem. The products sold by Microsoft to businesses and government offices were generally out of the price range of most homeowners.

In 1991, Linus Torvald tried filling the affordable software void by initiating the development of a free UNIX version for PCs using the Free Software Foundation's toolkit. His rapid success attracted many hackers on the Net—who gave him their very useful feedback on how to improve his product. Eventually, as a result of this tribal collaboration, Linux was developed, a complete UNIX available for free (meaning at no cost to users) and having redistributable sources.

In 1991, the PGP (Pretty Good Privacy) encryption program was released by Phillip Zimmermann, who later became involved in a complex three-year criminal investigation. The U.S. government alleged that the PGP program violated export restrictions for cryptographic software.

In 1992, the Michelangelo computer virus attracted a lot of news media attention because, according to U.S. computer security expert John McAfee (who had his own computer security software company), it was believed to cause great damage to data and networked computers. Those fears turned out to be greatly exaggerated, however, as the virus did little to the computers it invaded. Some say that McAfee "hyped" the damage the virus could cause in order to sell more of his software.

It was about this time that Jean Armour Polly coined the term *surfing the Net,* even though just over one hundred Websites existed on the Internet—and they were still very one-dimensional. Despite the shortage of Websites, though, in 1993 there were more than 1 million computers linked to the Internet.

In 1993, there were good events and bad events in terms of computer security issues. On the good side, White Hat Scott Chasin started BUGTRAQ, a full-disclosure online mailing list dedicated to issues about computer security—including software vulnerabilities (typically related to Microsoft products), methods of exploitation, and fixes for the discovered vulnerabilities. On the darker side, Randal Schwartz used the software program "Crack" at Intel Corporation for what he thought was "appropriate use" for cracking password files at work. He was later found guilty of illegal cracking activities under an Oregon computer crime law after Intel officials said that Randal had "crossed the line" into illegal and unauthorized activity. Schwartz was sentenced to five years' probation, ordered to complete almost 500 hours of community work, and was to pay Intel almost $70,000 in restitution.

In 1994, headlines were on fire with the story of cyber thieves led by Vladimir Levin. They cracked Citibank's computers and made transfers from customers' accounts without authorization totaling more than $10 million. Although in time Citibank recovered all but about $400,000 of the illegally transferred funds, that point was not featured by the news media. Levin got a three-year prison sentence.

The year 1994 also saw its creative geniuses. Two Stanford University students, David Filo and Jerry Yang, started their cyber-guide in a campus trailer as a way of tracking their interests on the Internet. The cyber-guide later became the popular search engine www.Yahoo.com (which means "Yet Another Hierarchical Officious Oracle").

Also, Canadian James Gosling, a native of Calgary, Alberta, headed a creative team at Sun Microsystems with the objective of developing a programming language that would change the simplistic, one-dimensional nature of the Web. The feat was accomplished, and the name given to the programming language, Java, was apparently coined in a coffee outlet near the group's Palo Alto, California, offices. Java allowed software on any operating system—the popular Microsoft Windows, Apple's Macintosh, or UNIX—to communicate by streaming bits of universally translatable data. The impact on the Internet was astounding: the "WorldWideWeb" became the "World Wide Web." In other words, the Web was no longer one-dimensional but allowed for a user-friendly, interactive environment where online users could leave each other e-mail messages, download songs, or order books—without leaving their homes or worksites (Brearton, 2005).

In 1995, political activism utilizing the Web became obvious, and with its popularity came a new online activity known as "hacktivism"—using the Web to send and receive politically motivated messages. For example, White Hat hacktivists squashed the U.S. government's highly controversial Clipper proposal, one that would have placed strong encryption (the process of scrambling information into something seemingly unintelligible) un-

der the government's control. Also White Hat in nature, the CyberAngels was founded in the United States, now the world's
oldest and largest online safety organization designed to track
and assist authorities in charging cyber-stalkers (those who stalk
others using the Web), cyber-harassers (those who harass and
threaten others using the Web), and cyber-pornographers (those
who make or distribute pornography, particularly child pornography, using the Web).

In 1995, the Apache Software Foundation, a nonprofit corporation, evolved after the Apache Group interacted. As a sidebar
update, this foundation eventually developed the now-popular
Apache HTTP Server, which runs on virtually all major operating systems. HTTP stands for "hypertext transfer protocol" and
is used to transfer data over the World Wide Web. All Internet
site addresses begin with http://. So, when a computer user
types a URL, or Uniformed Resource Locator, into the browser
(software application used to locate and display Web pages) and
hits the "Enter" key, the computer sends an HTTP request to the
correct Web server. The Web server, designed to handle such requests, then sends the user the requested HTML (or hypertext
markup language) page (SharpenedNet.com, 2006).

With the Web now available to businesses, 1995 saw the
launch of the first online bookstore by Jeffrey Bezos—the now
hugely popular international e-business chain selling books,
DVDs, and assorted items at Website http://www.Amazon.com.
The projected Net sales for 2005 were in the U.S. $8–9 billion
range (Stone, 2006). Commercially speaking, Microsoft released
Windows 95—and sold more than a million copies of it in less
than five days—with strong sales being an indicator of the growing demand for software by home computer users.

In 1996 the World Wide Web had grown to 16 million hosts,
and White Hat hacktivists continued to be active online. They
mobilized a broad online coalition to defeat the U.S. government's "Communications Decency Act"—thereby preventing
censorship on the then quite active Internet.

Also in 1996, the National Information Infrastructure Protection Act (NIIPA) was enacted in the United States to amend
the Computer Fraud and Abuse Act of 1984, thus enhancing its
powers. It was also in 1996 that the Child Pornography Prevention Act (CPPA) was passed in the United States to curb the creation and distribution of online child pornography. The 1996
Child Pornography Prevention Act extended the existing federal

criminal laws against child pornography to the new computer media. As part of the overhaul, the definition of "child pornography" was extended to include "morphed" or computer-generated images, as well as online material marketed as child pornography (Goodwin, 2001).

In the same year, XML 1.0, or Extensible Markup Language, was developed by an XML Working Group, including Tim Bray, currently director of Web Technology at Sun Microsystems. Originally known as the SGML Editorial Review Board, the Working Group was formed under the auspices of the World Wide Web Consortium (W3C) in 1996. The goals of XML were that it (W3C, 1998):

- be straightforwardly usable over the Internet
- support a wide variety of applications
- be easy to write programs that process XML documents
- be human-legible and reasonably clear
- have few optional features, ideally zero
- have a design that is formal and concise.

In 1997, ARIN, a for-profit organization, was started to assign Internet Protocol address space for North America, South America, sub-Saharan Africa, and the Caribbean. Also in that year, the DVD (Digital Versatile Disc) format was released. Consequently, DVD players were released for sale, appealing to many consumers because with this new software and hardware technology, they could watch popular movies at home rather than go to theaters.

In 1998, the main activities of the White Hat hacker labs became Linux development and the mainstreaming of the Internet. Many creative and entrepreneurial White Hats launched Internet Service Providers (ISPs), selling or giving online access to many around the world—and creating some of the world's wealthiest corporate leaders.

One of those corporate leaders was a Canadian named Tim Bray, a native of Vancouver, British Columbia. Bray helped to create the index known as Extensible Markup Language, or XML. The beauty of XML is that it made the popular online auction, eBay.com, possible. Prior to this language's development, the Internet was one jumbled reference book—which was interesting to delve into but which was one big, disorganized mess. The beauty of XML is that it allowed computer programmers to attach "tags," or universal codes, to distinguish, say, a business name from a telephone number. This development pushed electronic business (known as e-business) to the forefront, primarily because purchase orders could be universally read and properly routed. As a primary language of Web servers, XML propelled the activities of the estimated 1 billion Internet users in 2005. Equally as exciting, in 2005 e-Bay was estimated to have pumped $45 billion into the U.S. economy (Brearton, 2005).

Also in 1998, the Digital Millennium Copyright ACT (DMCA) was passed in the United States to implement certain worldwide copyright laws to cope with emerging digital technologies. By providing protection against the disabling or bypassing of technical measures designed to protect copyright, the DMCA encouraged owners of copyrighted works to make them available on the World Wide Web in digital format. Also during this year, the Boston hacker group L0pht testified before the U.S. Senate and warned about the many software vulnerabilities that could easily be exploited with widespread Internet use.

The year 1999 saw some incidents relating to harm to persons through Internet usage. One such cyber-criminal was nineteen-year-old Eric Burns, known on the Internet as Zyklon (a nerve gas). Besides cracking computers of U.S. companies, the White House, and the U.S. Senate, Burns also cyber-stalked a young woman named Crystal, one of his classmates. For these crimes, Burns eventually served fifteen months in federal prison, paid $36,240 in restitution, and was not allowed to touch a computer for three years after his release.

The year 1999 also witnessed some alleged incidents relating to harm to property as a result of Internet use. One such case involved a then fifteen-year-old Norwegian hacker known as "DVD-Jon." Acting with others in the Computer Underground, this young man developed a software program—called DeCSS—to circumvent the Content Scrambling protection on DVDs, thus making it possible for users to view DVDs with the open source Linux operating system. DVD-Jon then published his DeCSS decryption program on his Website so that other online users could download the decryption program and use it. In January 2000, the U.S. DVD Copy Control Association and the Norwegian Motion Picture Association, allies of the U.S. Motion Picture Association of America, filed charges against him in Norway. By January of 2003, DVD-Jon had been acquitted. The courts found that although DeCSS could be used to make illegal (that is, pirated) copies of the DVDs, the courts could not prove that this was DVD-Jon's motive for developing the software program or that others had used the program on illegally obtained DVD films (Vaagan and Koehler, 2005).

In the same year, the Internet was infected by the Melissa virus. It moved rapidly throughout computer systems in the United States and Europe. In the United States alone, the virus struck more than 1 million computers in 20 percent of the country's largest corporations. Months later, David Smith pleaded guilty to creating the Melissa virus, named after a Florida stripper. The virus is said to have caused more than $80 million in damages to computers worldwide.

Also in 1999, the Napster online music file-sharing system—used by Internet subscribers to copy and swap songs for free—began to gain popularity on Websites where students had access to high-speed Internet. Napster was developed by university students Shawn Fanning and Sean Parker, and it attracted more than 85 million online registered users before it was shut down in July 2001 as a violator of the Digital Millennium Copyright Act (DMCA).

And in that year, more legislation appeared. The Gramm-Leach-Bliley Act of 1999 was passed in the United States to provide limited privacy protection against the sale of individuals' private financial information. The intent of the act was to stop regulations preventing the merger of financial institutions and insurance companies. However, by removing these regulations, experts became concerned about the increased risks associated with financial institutions having unrestricted access to large databases of individuals' personal information—and the likelihood that this information could eventually find itself released for all to see on the Internet.

2000. In 2000, the emotional tone of high-tech moved from "positive" and "entrepreneurial" to "negative" and "fearful." Fears were running high, for example, about data losses resulting from Y2K (Year 2000) computer glitches. There were also extraordinary financial losses reported by investors wanting to earn major gains on the high-tech stock market—which had boomed in the 1990s during the growth of the Internet. There were also fears announced by North American government and security agencies of increasing numbers and types of cyber-crime, resulting from more widespread use of the World Wide

Web by individuals of all ages. Cyber-criminals no longer needed to be tech-savvy themselves. They could easily download cracking or defacing formulas from Websites—a common practice of "newbies" or "scriptkiddies" in the Computer Underground.

One high-profile cracking case making the news in 2000 was that of a fifteen-year-old Montreal, Canada, boy named "Mafiaboy." This case raised concerns in North America and around the globe about Internet security, following a series of Denial-of-Service attacks on several high-profile business Websites, including Amazon.com, eBay.com, and Yahoo.com. A year later, in January 2001, Mafiaboy admitted to cracking Internet servers and using them to start the cyber-attacks, causing legitimate online purchasers to be unable to access the business Websites to complete their transactions—and causing the businesses, in turn, to lose sales. For these exploits, Mafiaboy served just eight months in a youth prison and was fined $250. It was about this time that International Business Machines (IBM) experts estimated that online retailers could lose $10,000 or more in sales per minute if service was not available to customers because of such attacks.

In 2000, John Serabian, the CIA's information issue manager, said in written testimony to the U.S. Joint Economic Committee that the CIA was detecting with increasing frequency the appearance of government-sponsored cyber-warfare programs in other countries. Also in 2000, Dr. Dorothy Denning, a cyber-crime expert at Georgetown University, gave testimony before the U.S. Special Oversight Panel on Terrorism, affirming that cyberspace was constantly under assault, and that it was fertile ground for cyber-attacks targeted at individuals, companies, and governments. She warned that unless critical infrastructure computer systems were secured, conducting a computer operation that could bring physical harm to society's citizens on a wide scale could become as easy over the next few years as penetrating a Website and changing the words on it.

In 2000, some high-tech companies admitted that their companies' networks were vulnerable to attack. For example, Microsoft, the company of Bill Gates, admitted that its corporate network had been cracked, and that the source code for future Windows products had been seen. The suspected cracker was assumed to be from Russia.

Also in 2000, approximately 55,000 credit card numbers were stolen from Creditcards.com, a company that processes credit transactions for e-businesses. Almost half of these stolen credit card numbers were publicized on the World Wide Web when an extortion payment was not delivered by the company— a problem when it comes to online privacy (the state of being free from unauthorized access) and trust (having confidence in the company partner) by citizens making use of e-businesses.

2001. In 2001, computer worms and viruses crawled into the mainstream vocabulary because so many home and office computers were hit. For example, the Code Red worm compromised several hundred thousand computer systems around the globe in less than fourteen hours, overloading the Internet's capacity and costing about $2.6 billion in damages. The worm struck again in August 2001. Writer Carolyn Meinel noted in a piece in *Scientific American* that the worm was like a computer "disease" that had computer security researchers more worried than ever about the integrity of the Internet and the likelihood of imminent cyber-terrorist attacks. She likened the worm to electronic snakebites that infected Microsoft Internet Information Servers— the lifeline to many of the most popular Websites around the world.

Also in 2001, the Anna Kournikova virus was placed on the Net by Jan de Wit, also known as OnTheFly. The creator was a twenty-year-old from the Netherlands who was eventually arrested for this exploit and made to perform 150 hours of community service. Also in that year, NIMDA (ADMIN spelled backward) arrived, a blend of computer worm and virus. It lasted on the Web for days and attacked an estimated 86,000 networked computers. NIMDA demonstrated that some of the cyber-weapons available to organized and technically savvy cyber-criminals now had the capability to adapt to the local cyber environment.

And in 2001, there was excitement raised—as well as fears of unlawful arrest among White Hat hackers—at the DefCon 9 hacker convention in Las Vegas when the Russian Dmitry Sklyarov was arrested before speaking on the decryption software particulars that he had developed for his Russian employer, ElcomSoft. The software in question allowed users to download e-books on the Internet from supposedly secure Adobe software to more commonly used PDF computer files. The Electronic Frontier Foundation (EFF) argued that jurisdictional issues applied in this case and that Dmitry's behavior was perfectly legal in Russia, where he had conducted his acts. Eventually the courts agreed with the EFF, and Dmitry and his company were found not guilty (ibid.).

In 2001, a *Los Angeles Times* newspaper article spread fears about an imminent apocalypse caused by a computer. Crackers attacked a computer system controlling the distribution of electricity in California's power grid. Power was out for almost two weeks. According to the newspaper, the attack appeared to have originated from individuals in China's Guangdong province. The cyber-attack appeared to have been routed through China Telecom.

Extreme fear was raised on September 11, 2001, when terrorists attacked the World Trade Center and the Pentagon using commercial airlines. Hundreds of passengers were killed. For the first time in U.S. history, citizens became familiar with the intense fear generated by such political attacks.

On October 23, the USA Patriot Act of 2001 was introduced by James Sensenbrenner with the intent of deterring and punishing terrorist acts in the United States and enhancing law enforcement investigatory tools. The introduction of this act was a reaction to the September 11, 2001, terrorist attacks. Related bills included an earlier antiterrorism bill that passed the House on October 12, 2001, and the Financial Anti-Terrorism Act, which became law on October 26, 2001.

Internationally in 2001, group action was being taken to combat cyber-crime. On November 23, for example, the Council of Europe opened to ratify its newly drafted Convention on Cybercrime. The convention was signed by thirty-three states after the council recognized that many cyber-crimes could not be prosecuted by existing laws—which were typically local in jurisdiction. The convention was, indeed, the first global legislative attempt to set standards on the definition of cyber-crime and to develop policies and procedures to govern international cooperation to combat cyber-crime on the Internet.

On an optimistic note, in 2001 online gaming was becoming a very positive social force for the Internet. Online gaming, or Massively Multiplayer Online Role-Playing Game (MMORPG), is a form of computer entertainment played by one or more individuals using the Internet. Before online gaming, computer games were single-layer against a computer component. In contrast, online gamers can compete against or cooperate with human players through the Internet. The "I can see you, and you can see me" principle makes online gaming more interactive and exciting. According to a forecast made by DataMonitor.com in 2001, the global online gaming market would generate more than U.S. $3 billion in revenues and draw in over 113 million users by 2005.

It is interesting to note that this prediction appears to have been correct, for recent evidence shows that online gaming in Asia can be called the most successful software industry to date. As a case in point, the online game Lineage, developed by NC-Soft.com of Korea, made approximately U.S. $133 million in profits in 2002, and about U.S. $144 million the following year. Moreover, there were an estimated 2 million online gamers in South Korea in 2002 and more than 2 million players in Taiwan by 2003. The success of online gaming has shown huge growth potential for related industries, such as broadband network, on-line payment businesses, and Internet cafes. Online gaming has also, however, brought about the likelihood of online cheating and fraud (Chen et al., 2004).

2002. On July 10 and 11, 2002, U.S. Representative Richard Armey (R-TX) introduced a congressional bill known as the Homeland Security Act of 2002 to establish the Department of Homeland Security. The act was signed into public law by the president of the United States, and Section 225 of the act was known as the Cyber Security Enhancement Act of 2002; it was specially designed to keep U.S. cyberspace safe.

In 2002, the headlines reported a number of cyber-crimes committed by women, indicating that as far as gender equity goes, women can commit cyber-crimes just as well as men. For example, a seventeen-year-old female cracker from Belgium, known as Gigabyte, claimed to have written the first-ever virus in the programming language C# (pronounced C sharp), and a fifty-two-year-old Taiwanese woman named Lisa Chen pleaded guilty to pirating hundreds of thousands of software copies worth over $75 million. She was sentenced to federal prison in the United States for nine years—one of the most severe sentences ever given for such a crime.

In 2002, problems associated with online gaming started to be reported. For example, according to National Police Administration of Taiwan statistics, there were more than 3,553 cyber-crime cases in 2001, for which 3,983 cyber-criminals were prosecuted during 2002—an indicator that sometimes cyber-criminals work with others. Within these cases, more than 1,300 cases (37 percent) were related to online gaming, primarily online cheating (ibid.).

2003. In 2003, the notion of the vulnerability of wireless computer systems came to the forefront. For example, one interesting case involved a computer security analyst by the name of Stefan Puffer. After he figured out that the Harris County district clerk's wireless computer network was vulnerable to crack attacks, he warned the clerk's office that anyone with a wireless network card could gain access to their sensitive data. He was charged by police but was later acquitted by a Texas jury.

On April 30, 2003, stronger measures were passed in the United States to fight cyber-pornography. The PROTECT Act (Prosecutorial Remedies and Tools against the Exploitation of Children Today Act) of 2002 was signed into law by President George W. Bush, which not only implemented the Amber alert communication system to alert the nation about kidnapped children but also redefined child pornography to include images of children engaging in sexually explicit conduct with adults as well as computer-generated images indistinguishable from real children engaging in such sexual acts. Prior to the passage of the PROTECT Act, the definition of child pornography came from the 1996 Child Pornography Prevention Act, or CPPA.

In 2003, Paul Henry, vice president of an Internet security firm in Florida called CyberGuard Corporation, noted that cyber-security experts predicted that there was an 80 percent probability that a cyber-attack against critical infrastructures in the United States would occur as early as 2005. He said that the technology was available to those who wanted to use it—including organized criminals and terrorists. This fear was also manifested in a July 2003 poll of more than 1,000 U.S. adults conducted by the Pew Internet and American Life Project. The poll found that 58 percent of the female respondents and 47 percent of the male respondents feared an imminent national attack on critical infrastructures—brought on through a combination of cyber-attacks on networked computers running those critical infrastructures and on-ground terrorist attacks. Similar fears were raised when George Mason University doctoral student Sean Gorman's dissertation showed, using mathematical equations and charts, how terrorists could attack the communications network binding the United States.

More worms and viruses crawled onto the World Wide Web and wreaked havoc in August of 2003. The Blaster worm surfaced on August 11, exploiting security holes found in Microsoft Windows XP. The Welchia worm surfaced the same day, targeting active computers on the Internet. The latter went to Microsoft's Website, downloaded a program that fixes the Windows holes (known as a "do-gooder"), and then deleted itself. The most damaging of the three malware irritants was the e-mail–borne SoBigF virus, the fifth variant of a "bug" that originally invaded computers in January 2003 and then resurfaced on August 18. The damage caused by lost production for these three worms was reportedly in excess of $2 billion for just an eight-day period. It was about this time that John McAfee, the founder of the antivirus software company of the same name, claimed that there were more than 58,000 virus threats against Internet traffic, and that somewhere between ten and fifteen new online viruses and worms are discovered daily.

On August 14, 2003, fears of a cyber apocalypse were heightened for a period known as the Blackout of 2003, when the East Coast of the United States and the province of Ontario, Canada, were hit by a massive power blackout—in fact, the biggest ever affecting the United States. Some utility control system experts claimed that two events—the August Internet worm and virus invasions and the Blackout of 2003—might have been linked, because the Blaster worm, in particular, may have degraded the performance of several lines connecting critical data centers used by utility companies to control the power grid.

On September 15, 2003, the U.S. government, through the Department of Homeland Security, took even greater steps to tighten homeland security—including cyberspace—when they created the U.S.-Computer Emergency Response Team (or US-CERT). Moreover, technical groups such as the National High-Tech Crime Unit (NHTCU) in Britain began working with antivirus company experts to find patterns in the coding of some of the most destructive Internet worms and viruses to determine if they were the products of organized underground groups or other crime affiliates. NHTCU experts thought that perhaps hidden in the lines of code would be hints of the creators' identities, their motives, and maybe information about forthcoming cyber sabotage plans.

As a countermeasure against Internet crimes, in October 2003 an international consortium—including the U.S. Department of Homeland Security, the UK National Infrastructure Security Coordination Center, Canada's Office of Critical Infrastructure Protection and Emergency Preparedness (OCIPEP), and the U.S.-based SANS Institute (committed to fighting cyber-crime)—released a list of the top twenty known Internet security vulnerabilities. The consortium had as their objective the defining of an absolute minimum standard of security for networked computers around the globe.

In October 2003, large, wealthy Internet companies faced stiff penalties for infringing Intellectual Property Rights (IPR). For example, the Internet search giant Google was fined 75,000 Euros for letting marketers link their advertisements on the Net to trademarked search terms—a ruling that was said to be the first of the kind. At the time, the court gave Google a month to stop their illegal practices.

On November 6, 2003, the large, wealthy software companies started fighting back against cyber-criminals. For example, Microsoft took the very unusual step of creating a $5 million fund to track down malicious crackers targeting the Windows operating system. That fund, by the way, included a $500,000 reward for information leading to the arrest of the designers of Blaster and SoBigF. This Wild West–like bounty underscored the perceived "problem" environment of the Internet frontier. Some cynical security critics said that this reward had more to do with saving the company's public relations than it had to do with cyber-crime prevention.

Without question, 2003 was a year of spam-banning as well. In November, the U.S. Federal Trade Commission (FTC) created a national spam database and encouraged networked citizens to forward to them all the e-mail spam they received. Unsolicited, unwanted, impersonal e-mail received through the World Wide Web is considered to fall into the general heading of "spam." The FTC noted that in 2002, informants had taken it upon themselves to report more than 17 million complaints about spam messages to federal agents for investigation. In fact, the FTC said that they received nearly 110,000 complaints daily.

As a spam countermeasure, in November 2003 the U.S. Senate passed the CAN-SPAM Act of 2003, formally known as the Controlling the Assault of Non-Solicited Pornography and Marketing Act of 2003. Its purpose, as its name implies, was to regulate interstate commerce by placing limitations and penalties on the transmission of spam through the Internet. Penalties that could be imposed were fines as high as U.S. $1 million and imprisonment for not more than five years if found guilty of act infringement.

2004. In 2004, large companies started pursuing U.S. citizens who swapped songs using the Internet without paying royalties to the song artists. On January 21, the Recording Industry Association of America (RIAA), for example, said that it had identified 532 song swappers through the cyber trails that their computers left when the swappers illegally downloaded music from the Internet. The swappers, identified by their Internet Protocol (IP) addresses, were targeted in four lawsuits in New York and in Washington, D.C. The lawsuits were filed using the so-called John Doe process, allowing the RIAA to sue defendants whose names were not yet known.

More worm releases emerged in 2004. For example, on January 26, the worm W32.Novarg.A@mm, also known as MyDoom, spread like wildfire through the World Wide Web, arriving as an attachment with the file extension .bat ,.cmd., .exe, .pof, .scr, or .zip. It affected Windows 2000, Windows 95, Windows 98, Windows Server 2003, and Windows XP systems. It did not, however, affect DOS, Linux, Macintosh, OS/2, UNIX, or Windows 3.x systems. Clearly, the Microsoft Corporation was the premeditated target. The damage done by MyDoom was estimated to be U.S. $2 billion worldwide.

Although much of the Internet development headlines have been increasingly negative since 2000, the occasional bright light does appear. For example, in mid-August of 2004, would-be high-tech billionaire White Hats by the names of Sergey Brin and Larry Page began their long-awaited Initial Public Offering (IPO) on the stock market of their Internet e-business Google—home of the world's leading Internet search engine.

It is interesting to note that in September 2004, the *Computer Industry Almanac* estimated that the worldwide number of Internet users would top 1 billion, with the United States alone accounting for 185 million users. The *Almanac* also noted that there will be little Internet user growth in the developed countries over the next five years, but that many users will be supplementing their PC Internet use with Smartphone and mobile device Internet applications. Elsewhere around the globe Internet use is growing strongly in China, which actually had greater Internet growth than Japan in 2003. The *Almanac* also predicted that growth in Internet users will continue in developing countries for the next decade (Computer Industry Almanac, 2004).

On the international scene in 2004, Chinese authorities, worried that some of the material appearing on the Internet was immoral or violent in nature, closed more than 12,000 Internet cafes where Chinese citizens of all ages connected to the World Wide Web.

2005. In January 2005, under public pressure from U.S. citizens, the FBI (Federal Bureau of Investigation) abandoned its custom-built and highly controversial Internet surveillance technology known as Carnivore, designed to read e-mail and online communications of suspected terrorists and cyber-terrorists. The FBI moved to using commercially available software and encouraged Internet providers to conduct wiretaps on suspected individuals, and to forward the findings to government agents.

On January 21, 2005, in an embarrassing Harvard University incident, the confidential drug purchase histories of Harvard students and staff and the e-mail addresses of undergraduates (guaranteed by university officials of nondisclosure) appeared on Harvard's iCommons Pool Tool. To make matters worse, this personal health history information was accessible for months to anyone having an Internet e-mail account and a few minutes' time to investigate the particulars of anyone having an eight-digit ID number at the university. This vulnerability in Harvard's computer network underscored the difficulty of securing the system when there is prevalent use of ID numbers to verify individuals' identities.

In February 2005, the National Technology Readiness Survey (NTRS) results indicated that spam, besides being very annoying, costs nearly U.S. $22 billion annually in lost productivity in the United States alone. That same month, while delivering a speech to security experts at the RSA Conference in San Jose, California, Microsoft Corporation founder Bill Gates said that his company will give away software to combat privacy-invasive cyber nuisances like spyware and adware. Spyware is covert software whose mission it is to capture data about online users' Internet-surfing habits. Adware, a form of spyware, gathers information on unsuspecting users with e-mail pop-up ads or other marketing inducers.

In February 2005, a report released by a U.S. legislative committee found that information on the Websites of the New York

Motor Vehicles, the New York Department of Education, the New York Department of Correctional Services, the State Division of Military and Naval Affairs, and the New York Power Authority had been defaced seventy-two times from 1999 to early December 2004. The chair of the oversight committee said that because state and private companies are not keeping important personal and homeland security online information safe, identity thefts (stealing the personal details of another person, such as social security numbers or credit card numbers and using them for one's own benefit)—or worse—could occur.

In February 2005, another warning came, from the vice president of security products for the OpenService company. He warned that from the legal requirements of the Sarbanes-Oxley Act, recently passed in the United States—an act intended to prevent further financial scandals like Worldcom—any breach in IT security poses a risk to a company's internal information system. Since IT underlies the critical business of recording and reporting all financial activity, a lack of control over IT security would imply a lack of control by a company over its financial reports—a direct violation of Section 404 of the act, in particular.

Other computer system vulnerabilities made headlines in 2005. One of those occurred on March 2 when Harvard Business School announced that they would reject 119 applicants who followed a cracker's instructions for peeking into the school's admission Internet site to see if they had been accepted—one month before the acceptances would be made public. The next day, MIT's Sloan School of Management said that they would do the same with crackers of their Internet site. An anonymous person known as "brookbond," who said that he was a male who specializes in IT security, posted the how-to-crack instructions on Business Week Online's technology forum just after midnight on March 2, making Harvard University's Website vulnerable to attack for more than nine hours.

Other kinds of aggressive Internet-related incidents were reported in March 2005. For example, on March 5, a cyber-war broke out between Indonesia and Malaysia, instigated by a dispute over the Ambalat oil fields in the Sulawesi Sea—and, perhaps, the termination of amnesty for illegal Indonesian workers. Kuala Lampur officials were upset over an intrusion into its waters by an Indonesian naval vessel after Indonesian president Bambang Yudhoyona ordered the military to show itself in the Sulawesi Sea jurisdiction under dispute. Within twenty-four hours of that act, the Website of Universiti Sains Malaysia was cracked and plagued with aggressive anti-Malaysian messages with an Indonesian twist.

Two days later, on March 7, 2005, documents seized from three members of the Lashkar-e-Toiba (LeT) terrorist group killed in an encounter with Indian police indicated that they had planned to execute a "suicide crack attack" on the computer networks of companies having software and computer chip design outlets in Banglahore and Karnal Singh. These companies would include such readily recognizable names as Intel, Texas Instruments, and IBM. The motive behind this act was LeT's demanding independence for the Indian states of Jammu and Kashmir. These cyber-terrorists said that they wanted to attack the high-tech companies to hinder the economic engine of India. The companies who were targeted said that not only do they have tight entry requirements in their facilities in those locations but they also have well-designed disaster recovery plans in place.

In the ever-evolving world of wireless technology and the Internet, more opportunities for virus creation have emerged. For example, on March 8, 2005, data security firms said that a new virus capable of attacking users' cellular phones had been developed. F-Secure Corporation, a Finnish software security company, said that the Commwarrior virus appeared to be the first one capable of spreading through online multimedia messaging services containing photos, sound, or video clips. Two days after this announcement, Trend Micro, another IT security firm, issued an alert regarding two more worms that spread through users' MSN Messenger, a widely employed instant messaging platform found in particular in the Asia Pacific region. Known as kelvir.b and fatso.a, these worms were reported invading the cyberspace in both the United States and the Asia Pacific region.

Despite the heavy negative tone that has pervaded the cyber community since 2000, there have been some cyber-incidents reported by the news media that could evoke a chuckle or two. One of these occurred on or about March 10, 2005, when the press reported that Limp Bizkit singer Fred Durst had taken legal action against ten Website operators for posting his self-created sex tape after his computer was cracked.

Fears about the vulnerability of wireless computer networks continued to make the news in 2005. On or about March 10, according to a study commissioned by RSA Security, the huge growth in the use of wireless networks by businesses around the globe is making those networks increasingly more subject to drive-by cracking—meaning that someone can literally drive by a company and connect to the wireless network to do online Google searches, say, or to view online pornography (Grami and Schell, 2004). As a case in point, in Europe's financial districts, the wireless networks seem to be growing at an annual rate approaching 70 percent. What is really worrisome is that more than a third of the businesses making use of this technology are unprotected from crack attacks. By comparison, noted RSA Security, almost 40 percent of the businesses in New York seem to be unprepared for such exploits, and more than a third of the businesses in San Francisco seem to be unprepared.

Moreover, in March 2005, evidence surfaced of the poorly regulated Internet information brokerage industry. On or about March 10, cyber-criminals used passwords stolen from legitimate users to access the LexisNexis Group's data on as many as 32,000 Americans. Similar computer network breaches occurred at ChoicePoint, Inc., and at the Bank of America. These intrusions prompted calls for U.S. federal government oversight through the General Services Administration (GSA) to look into the matter. The GSA agreed to review the online security policies of the Bank of America and four other SmartPay contractors providing credit avenues for online transactions—Bank One of Delaware, Citibank of New York, Mellon Bank of Pittsburgh, and the U.S. Bank of Minneapolis. Recommendations were to follow about providing adequate protection for the safety of online personal information.

Elsewhere around the globe, Denial-of-Service (DoS) crack attacks made the headlines yet again. On March 16, for example, five crackers in The Netherlands were found guilty of disabling a number of Dutch government Websites because of DoS attacks a year earlier. Citing protests against recent cabinet proposals as the motive for their attacks, the hacktivist group went by the moniker Oxlfe Crew. The spokesperson for the group, an eighteen-year-old who said that he would appeal his thirty-eight-day detention sentence, argued that there was no evidence to prove that he was involved in the Internet attacks. This was the first time that anyone in The Netherlands had been convicted of such a cyber-crime.

News releases in 2005 from Internet security firms continued to report significant growth in the number of attacks made on North American companies' computer networks. For example, on March 21, 2005, Symantec Corporation issued a report saying that Internet attacks had grown by 28 percent in the second half of 2004, compared with the first six months. On average, Symantec noted, businesses and other agencies received about fourteen attacks on their computer networks daily in the second half of 2004, relative to about eleven attacks in the first half of the year. Moreover, crackers increasingly turned to mobile computers to exploit networks—using such online tools as phishing, spyware, and adware. Phishing is a form of identity theft whereby a scammer uses an authentic-looking e-mail from a large corporation to trick e-mail receivers into disclosing sensitive personal information online, such as credit card numbers or bank account codes.

Another release at about the same time by a security company called Mazu Networks indicated that almost half of the 230 companies surveyed had had a computer worm outbreak in 2004, despite their increased spending for security. Although about three-quarters of the IT professional respondents said that their companies had increased security spending in 2004 to comply with the Sarbanes-Oxley Act, less than 15 percent of the respondents reported that they were confident that their networks could stop cyber-crime.

Despite the accumulating negative news in 2005, there was the occasional positive word. For example, on April 8, 2005, in a landmark legal case, spammer Jeremy Jaynes of North Carolina, described by prosecutors as being among the top ten spammers in the world, was sentenced to nine years in federal prison. This was the first successful felony prosecution in the United States for transmitting spam over the Internet.

Also in 2005, some advances in Internet technology sparked thoughts of cyber-crime relief. For example, during the week of April 25, Bill Gates said that his company planned to put hardware computer security into a silicon chip, instead of relying only on software in the next release of the Windows PC operating system—which would be available sometime in 2006. Also, on May 23, Apple Computer, Inc., announced that they were in talks with the Intel Corporation about using their Intel chip in the Macintosh computer line. Using Intel chips in the Macintosh would make it feasible to install the hugely popular Windows software on their Mac computers. With that announcement, not surprisingly, Apple shares rose on the Nasdaq stock market.

There were also technological advances in 2005 that contributed to medical benefits. For example, the U.S. Food and Drug Administration (FDA) had already approved for use Implantable Cardioverter-Defibrillators, or ICDs, made to transmit a patient's heart-monitoring data (including electrocardiograms) over telephone lines. Although ICDs are meant to assist doctors in monitoring their patients' heart conditions from locations other than at the doctor's office, there have already been IT security concerns voiced by security experts. The security concern lies around the remote relaying system, whereby patients hold a wand above their chest. The information is then sent over the telephone line to the doctor in encrypted form. Although the FDA has not yet approved physicians being able to adjust the defibrillator over the phone, the technology does allow for that. The fear, then, is that some ill-motivated Black Hat cracker will attempt to adjust the defibrillator of some targeted individual by cracking telephone lines or by altering the information relayed to the physician (Adler, 2005).

Also in 2005, there were indications that many North American cyber-citizens had become addicted to using the Internet for a variety of reasons—reading e-mail, surfing the Net for topics of interest to them, or purchasing consumer goods online. For example, during the first week of May 2005, Websense's annual Web@Work survey results indicated that more than 50 percent of the employee respondents said that they would rather forgo their morning coffee than lose their ability to surf non-work-related Websites during the workday. Moreover, according to a 2005 report released by Forrester Research, Inc., U.S. households now spend at least 30 percent of their media time online, and according to the Interactive Advertising Bureau, online ad revenues grow at a rate at or exceeding 33 percent per quarter.

In 2005 cyber-citizens outside of North America were also becoming addicted to the pleasures of the Internet. For example, in early August the newly released Chinese search engine Baidu.com stock soared over 350 percent on the first day of trading. The thirty-seven-year-old Chinese founder, Robin Li, told reporters that Chinese Internet use was still in its infancy, but that like North American search engine founders, he had hopes of becoming another high-tech billionaire, because Chinese citizens enjoy staying "connected" with the rest of the world through the World Wide Web. Although Li's statement contains much truth, on October 5, 2005, Chinese authorities shut down another two Internet sites popular among academics, journalists, and civil rights activists. One Website served as an online discussion group to report on anticorruption protests in a southern village around Beijing. The other Website serviced ethnic Mongolians.

In the fall of 2005, while many students in North America were returning to school, another young cracker was going to jail. On September 8, 2005, a Massachusetts juvenile pleaded guilty to cracking Paris Hilton's T-Mobile cellular phone and dumping personal information of hers on the Internet. He also admitted to cracking other computer systems in 2004, to stealing personal information, and to sending bomb threats through the Internet to high schools in Florida and Massachusetts. All of the incidents occurred in a fifteen-month period, beginning in March 2004. The harm to persons and to property over this short period cost an estimated U.S. $1 million, and like other young crackers before him, the youth was sentenced to eleven months in a detention center.

On a positive year-end note, 2005 ended with some happy stories. Although almost a quarter of the world's spam in the last three months of 2005 was sent from U.S.-based computers, according to UK antivirus company Sophos, the latest figures indicate for the first time that the United States accounts for one-quarter of all relayed spam—a significant drop from about 42 percent in 2004. The reduction is linked to the effectiveness of the CAN SPAM Act and to the courts' imposing harsh prison sentences and stiff financial penalties for those found guilty of the crime. In 2005 the U.S. spam offenders were followed closely by Chinese spam offenders, owning about 22 percent of the problem. Korea was the next spam offender, coming in with about 10 percent of the relayed spam (Evers, 2006).

If Robin Li needs good role models, he has found them in Google's founders Sergey Brin and Larry Page. With the news on November 17, 2005, that stock market shares of online search giant Google, Inc., had crossed the amazing $400-per-share mark, the cyber and financial worlds may have seen the birth of a new financial planetary system.

As of November 22, 2005, Google has a market cap of U.S. $120 billion—double the market cap of its nearest competitor, Yahoo, capped at U.S. $60 billion. For comparison, online auction house eBay has a market cap of U.S. $65 billion, and online publisher Time Warner has a market cap of U.S. $85 billion. There is little question that the success of Google is altering in a very positive way the tech industry's behavior—and is sending a strong message that tech billionaires can still be made on the Internet highway. In fact, Google's stock, with a price-earning ratio of 70, represents one of the richest deal-making currencies in the world. So, where are Brin and Page looking to invest their dollars? Although for a while they were interested in Web phone Skype, they have since backed away. To date their largest deal, made in 2003, seems to be the U.S. $102 million they paid for online advertisement upstart Applied Semantics, Inc. Today, Brin and Page are looking to spread their online wings in China.

2006. The New Year ushered in more cracking headline stories. For example, in January 2006, a twenty-year-old California cracker named Jeanson James Ancheta admitted to seizing control of hundreds of thousands of Internet-connected computers by making use of a zombie network to dump pop-up ads on targeted computers. He also rented the network to others so that they could mount cyber-attacks on selected Websites and send out spam. Ancheta cracked computers at two U.S. military Websites, earning himself about $60,000. Lawyers arguing for the government cited this case as the first to obtain significant profits from the use of botnets—large numbers of computers commandeered by crackers and marshaled for personal gain (MercuryNews.com, 2006).

In recent years, botnets have been used to overwhelm Websites with streams of data and feed off vulnerabilities in computers running the Microsoft Windows operating system. The vulnerable machines are generally those whose owners have not installed security patches, and zombie machine owners are unaware that programs have been installed on their computers that are being controlled by crackers in a remote fashion. The court documents apparently indicate that Jeanson would spend about $600 for expensive clothes and BMW car parts. He faces up to six months in prison and will have to pay the government restitution. At about the time of his arrest, the FBI released their latest findings indicating that viruses, worms, and Trojan horse software programs like the ones that Ancheta used cost U.S. companies U.S. $11.9 billion annually. Named after the Trojan horse of ancient Greek history, a Trojan is a particular kind of network software application developed to stay hidden on the computer where it has been installed. Like worms, Trojans generally serve malicious purposes, and today Trojans access the target computer through the Internet. Since Trojans can launch Denial-of-Service attacks, security experts suggest using antivirus software to protect networks (ibid.).

A more recent legal case regarding Intellectual Property Rights and Internet technology made the headlines in 2006. This fascinating case involved two high-tech companies: RIM, Ltd. (Research in Motion, Ltd., a Canadian company based in Waterloo, Ontario), and NTP, Inc. (a small U.S. company based in Virginia). Each company claimed that it owned the Intellectual Property Rights (IPR) and patents for the wireless handhelds used by business and government people around the world to read and respond to their e-mail while away from their computers. Named by RIM as "The BlackBerry," these nifty little handhelds have been given the name "crackberries" for their addicting ease of use. It is interesting to note that the devices are so popular that the U.S. federal government had requested that employees with mission-critical roles be exempted from an injunction, should one be levied by the courts (Arnone, 2006). However, the case ended in a negotiated settlement in March 2006. The CEO of RIM, Jim Balsillie agreed to pay NTP a one-time payment of $612.5 million, settling the dispute. Under this agreement, RIM received a license to NTP's patents going forward.

Other interesting news surfacing in 2006 included the U.S. partnership with developing nations in the war against cyberterrorism. For example, in January 2006, just over twenty officers from Indian law enforcement agencies were trained in New Delhi in a cyber Incident Response Course, sponsored by the U.S. State Department's Anti-Terrorism Assistance Program. Those attending the course were trained in methods of securing and preserving evidence at a cyber-terrorism crime scene. The participants also learned how to identify and seize digital evidence without altering its contents—an important issue for courts needing untouched evidence (NewKerala.com, 2006).

Moreover, the U.S. government was still concerned about the vulnerability of the computer networks responsible for critical infrastructures. In January 2006, officials in the Homeland Security Department said that it will maintain the confidentiality of critical infrastructure information submitted to the National Asset Database. Also, the department said that they will scrutinize all requests to view this database and will grant access only to select Department of Homeland Security employees and others only on a tightly controlled, need-to-know basis (Lipowicz, 2006).

The UK government is also concerned about tight control but in a different online context. That government has repeatedly expressed a desire to use e-enabled, or a Remote Electronic Voting (REV) system, in a general election to be held after June 2006—the latest date for the next election. From a security standpoint, say security experts, most existing REV schemes use some form of cryptography, either to secure the transmission of unchanged votes or to model some positive feature of public elections. REV schemes have also been proposed for nonelection purposes, such as for jury voting or for parliamentary voting, where it may be desirable to reveal the association between a voter and a vote. It will be interesting to see how this creative use of the modern-day Internet will turn out, for most present-day elections use a combination of in-person voting, regular mail voting, and small-scale online voting (Storer and Duncan, 2005).

On February 1, 2006, RIM, Ltd., secured a key victory after U.S. patent authorities rejected the last of five contested patents at the heart of the dispute, and immediately RIM shares rose nearly 9 percent on the Toronto Stock Exchange, as investors bet that RIM might be gaining the edge in the five-year patent saga. The patent agency's decision, although not final, is quite unlikely to be changed. However, NTP has thirty days in which to respond, after which a final court ruling is delivered. Either side can appeal the decision to a patent appeals board, the U.S. Court of Appeals for the Federal Circuit, and the Supreme Court. This case clearly shows how cumbersome and costly patent disputes can become. The patent rejection could also strengthen RIM's hand in continuing talks with NTP or in reaching a possible settlement—which is likely to top U.S. $1 billion (McKenna, 2006).

How the Internet Works

The General Nature of the Internet

Until 1996, few North American laypersons or lawyers understood how the Internet, or the World Wide Web, actually works. Then in 1996 and 1997, after the U.S. government passed the

Communications Decency Act (CDA) to regulate obscene information on the Internet, some lawsuits were filed challenging the CDA's constitutionality. One of these was the 1997 case *Reno v. ACLU* (that is, Reno, the U.S. attorney general, versus the American Civil Liberties Union). Details on the workings of the Internet using lay language were consequently reported in the court decisions pertaining to these cases (Findlaw, 2006).

Without question, even from the beginning the Internet was designed to be a huge network interconnecting many smaller groups of linked computers. The Internet never had a physical or tangible entry, as a door on a house provides entry to the rest of the house. Moreover, the Internet, even in 1969, was designed to be a decentralized, self-maintaining group of redundant links between computers and computer networks. The beauty of the Internet was that it was able to send data rapidly without human intervention and with the unique capability to reroute data if one or more links were damaged or unavailable. Therefore, from a military perspective, even if a portion of the network were destroyed during a war, with a redundant system, critical data could still reach its intended destination.

It is important to note that data travel along networks not as "whole" information but as very small information packets. In fact, even in ARPAnet's early days, the Internet used packet switching protocols to allow messages to be subdivided into smaller packets of information. These smaller packets would be sent rather independently to the required destination and be automatically reassembled by the destination computer. Although packets containing information for a larger message often travel along the same path to get to the final destination, some of them could be rerouted along a different path if one or more routers along the route became overloaded.

The reason that the networked computers can communicate with one another is as a result of a common communication protocol called Internet Protocol. All applications utilized on the Net are designed to make use of this IP—both in earlier days and today.

Gaining Access

Inasmuch as the Internet is not tangible, is not stored in any one place, and has no door in the true sense, how do Internet users gain access? There are a number of ways that individuals can gain access, but one of the earliest means was through a computer network linked directly to the Internet at an educational institution, business, or government office. The IT department would issue account numbers and passwords to legitimate users.

As more and more citizens bought PCs during and after the 1990s, individuals would more commonly pay a fee to an Internet Service Provider (ISP) and then receive a software package, a username, a password, and an access phone number. Equipped with a modem, the client could then log on to the Internet, browse the World Wide Web, or send and receive e-mail. Lately, ISPs have been offering high-speed services using DSL (a general term for any local network loop digital in nature) or cable-modem technology. ISPs are connected to each other through Network Access Points.

Today, laypersons often gain access to the Internet through commercial online services—such as America Online (AOL), CompuServe, and the Microsoft Network—which offer not only Internet access but also other useful content, such as news, entertainment, and lifestyle, diet, and marketplace information.

Ways to Communicate

The most common ways to communicate with others over the Internet include

- E-mail, or Electronic Mail
- Listserv
- Newsgroups
- Internet Relay Chat (IRC)
- Telnet
- Remote Retrieval

E-mail. Using e-mail, an individual can send a message to one or more other individuals also connected to the Internet. Unencrypted e-mail is not secure compared with regular posted mail (often called "snail mail" because of its comparatively slow delivery time), which is delivered in a sealed envelope via the government's postal service. Contents may be accessed, viewed, and altered by someone with computer access along the path from sender to receiver. It is quite common for online users to check on the receipt of e-mail by sending a follow-up e-mail if a response is not received in a timely fashion.

Listserv.   Probably the best way to think about listserv is to picture it as an online subscription to some topic of interest. Enlisted online users can then receive messages posted by other enlisted members and post messages for others on the listserv to receive. The kind of software that manages listservs is known as a mail exploder; it is a program that runs on the server where the list resides. It is this software program that gives instructions to those wanting to unsubscribe, and it does so without human intervention. Some listservs have open access, while others have more restricted access. Moderated listservs have someone who actually approves of the messages posted before sending them on to registered users.

Newsgroups.   Users connected to the Internet can also get articles posted to thousands of discussion groups called newsgroups. The articles are topic-arranged and distributed through an electronic bulletin board system called Usenet. When a user having access to a Usenet server (that is, a computer linked to the Usenet system) posts an article to a newsgroup, the server forwards the article to adjacent Usenet servers, making the article available on all Usenet sites with access to that newsgroup. Some Usenet newsgroups are moderated, while others are not. Once a message reaches a particular Usenet site, it is stored there temporarily. Users running software called "newsreader" can then sort articles by header information and respond to the information.

Internet Relay Chat (IRC).   Servers running software known as Internet Relay Chat allow multiple users "to talk online" and in real time by choosing one of many activated discussion channels. What this means is that some message typed on one user's computer will appear almost immediately on another user's computer monitor. There are a number of commercial online service providers from which users may partake of these online chats, including America Online and CompuServe. To keep their identities unknown to those in the chat rooms, users often go online under some contrived moniker, such as Mafiaboy. Although online chat rooms form an important communications link for citizens around the globe, many children and women have filed complaints to the authorities after being cyber-harassed or cyber-stalked in these virtual rooms.

Telnet. Technically speaking, Telnet is a terminal emulation program, or a program based on that protocol, which allows users to log on to the Internet. Stated simply, Telnet is an Internet application permitting a user's PC to act as a terminal to a remote system. For many users connecting to the Net, Telnet is not used, because it requires familiarity with the UNIX operating system. For example, a professor at one university can use Telnet to make use of the extraordinary computing power of a supercomputer located at some other university. One case in point is Canada's Shared Hierarchical Academic Research Computing Network (SHARCNET), having a network capacity of 6,000 processors and linking eleven leading Ontario academic institutions. Its purpose is to meet the rapidly growing high-performance computing demands of the twenty-first century and to advance research progress in the Canadian research community (Hewlett Packard Development Company, 2005).

Remote Retrieval. A common use of the Internet these days is to search for and retrieve information located on remote computers. There are basically three ways in which this search and retrieval process may be accomplished. One way is to use the File Transfer Protocol to transfer files between systems over the network, particularly from a host (that is, server) to a remote computer (that is, client). It is important to note that commercial browsers like Netscape have FTP capabilities built in.

The gopher protocol, a distributed document search-and-find network protocol, was released in 1991 by Paul Lindner and Mark McCahill. The gopher's original design for sharing documents was similar to that of the World Wide Web—and, as such, it has been replaced by the Web. Because the protocol had some features not supported by the Web, some IT experts consider it to have had a better protocol for searching and storing large data repositories. In fact, when the World Wide Web was first introduced in 1991, gopher was still popular. Then, in February 1993, when the University of Minnesota—which is where Lindner and McCahill attended university—announced that it would begin to charge users licensing fees to use the protocol, it underwent a huge decrease in popularity. Some security experts believe that gopher's downfall was brought about by its structure, which is limited compared with free-form hypertext formatting language, or HTML.

The third method to retrieve information from the World Wide Web is to use HTML. Today, programs browsing the Web can display HTML documents having text, images, sound, and video, and they can do so anywhere in the world. Although information on the Web is stored in individual HTTP (hypertext transfer protocol) servers, the connection of these computers to the Internet through W3C protocols (those used to exchange information on the Web) permits the linked information to become part of a single body of information. In short, the World Wide Web, or WWW, describes the network of HTTP servers using hypertext links both to find and to access files. That is why today all Internet site addresses begin with http://. Without question, the World Wide Web is the most advanced information system of an Internet nature and has embraced within its data model most information in earlier networked information systems using FTP, Usenet, and gopher.

Information contained on the World Wide Web is stored in many formats, as noted—video, text, images, and sounds, with each document having an address consisting of numbers. Moreover, most Web documents contain links—words or images—referring to other documents. Thus, when an Internet user clicks on any linked text or image (typically blue and underlined when displayed), results in the referenced document will be automatically displayed—regardless of where it is being stored around the globe. Alternatively, if a user types a URL (Universal Resource locator) into the browser and hits the "Enter" key, the computer will send an HTTP request to the correct Web server. The Web server, developed to handle such requests, then sends the user the requested HTML page.

Publishing on the World Wide Web

Today, hospitals, research centers, financial institutions, businesses, and universities have their information available on the World Wide Web. If information is available, it is said to be published. Getting published on the Web requires only a computer connection to the Internet and running server software on the computer that meets the W3C (World Wide Web Consortium) formatting standards. It is the meeting of such standards that allows connected computers to communicate and exchange information with one another.

The W3C was founded by Tim Berners-Lee, who once said that the World Wide Web is actually a universe of network-accessible data—an embodiment, as it were, of human knowledge. Today, the W3C exists to help the Web reach its fullest potential. It is a consortium of industry leaders wanting to promote standards for the Web's continued development and for greater interoperability between WWW products.

The W3C receives funding from its industrial partners to produce reference software and specifications. Despite this funding arrangement, the W3C upholds its claim not only that it is vendor-neutral but also that its products are free for everyone. Moreover, the W3C is international in character. Joint hosts include the MIT Laboratory for Computer Science (where Tim Berners-Lee is a professor) and the INRIA (the French National Institute for Research in Computer Science and Control) in Europe. Throughout their six research units in France (Bordeau-Lille-Saclay, Grenoble, Nancy, Rennes, Rocquencourt, and Sophia Antipolis), INRIA has more than 3,500 employees (Berners-Lee, 2006; INRIA, 2006).

Once published on the Web, publishers may choose to make their Websites open to the general pool of Internet users, or to close them by permitting access only to users with advanced authorization. If publishers want to limit access, they assign specific user names and passwords as a prerequisite for obtaining access to the published Website. If a Website is designed primarily for an organization's internal use, it is known as an Intranet. Thus, access to information is permitted only from other computers within the organization's local network.

Creating a Website and Getting a Domain Name Address

Creating a Website.   In the late 1990s, creating a Website used to cost anywhere from U.S. $1,000 to U.S. $15,000, with monthly operating costs varying according to the Website's purpose and the amount of traffic. Today, creating a Website can be done for free, if using, for example, http://www.freewebs.com, and it includes free bandwidth—up to 1.5 GB. More bandwidth—up to 50 GB—can be purchased for as little as $30 per month. Also, a domain name address can be registered on Websites like http://www .Yahoo.com for as little as $2.99 per year. The most popular domain name extensions include .com (meaning commercial) and .edu (meaning an educational institution).

Getting a Domain Name Address. The Domain Name System, or DNS, is a hierarchical system of naming hosts and placing the hosts into categories. The DNS is a way of translating numerical Internet addresses into word strings to denote user names and locations. For example, the Domain Name System http://rs .internic.net/ is also known as 198.41.0.13.

Any machine on the Internet has its own address, called the Internet Protocol address (IP address). The IP address looks something like this: 123.123.123.123—four numerical segments separated by dots. Any computer is reachable through its IP address. Because users cannot remember these numerical strings of IP addresses, an alternative, more user-friendly system was needed. For this reason, IP addresses were translated into more logical strings for humans to remember. The IP address for the Yale University computer science department, for example, is denoted as www.cs.yale.edu.

During ARPAnet's early development, one file called host.txt listed all of the organization's IP addresses. At the end of each day, all computers connected to the Internet would get the list from the Website where it was kept. With time, the number of connected hosts increased to such a degree that the size of the host file was huge, and the system was inefficient. Thus the DNS was invented—a hierarchical domain-based structure in which the Internet is divided into pieces called "domains." The pieces are categorized as top-level domains and subdomains. The top-level domains include generic and country domains.

The generic domains are .com (a commercial enterprise), .edu (an educational institution), .gov (a governmental agency), .int (an international institution), .mil (a military institution), .net (a network institution), and .org (a nonprofit organization). The country domains, allocated one per country, look like this: .au for Australia, .us for the United States, and .ca for Canada.

Each top-level domain is divided into several subdomains, with each domain having control over its subdomains. For example, the .edu domain covers all of the educational institutions and subdomains—such as Yale University, Princeton University, Rutgers University, and Harvard University. Moreover, the country domains have subdomains. For example, the .uk (the United Kingdom) and the .jp (Japan) domains have two common subdomains: .ac (standing for academic entity) and .com (standing for commercial entity). Each domain has a particular server with a table containing all IP addresses and domain names belonging to its domain.

An organization called the Internic maintains a database having all registered domains for the world. Anyone can query its database by means of whois. Although several organizations maintain whois databases, the Internic has the main database. So any company, institution, or organization wanting to have its own domain name must register it with Internic.

A number of whois servers exist around the globe. For example, in Amsterdam, there is the European whois server at RIPE (Reseaux Ip Europeans).

Navigating the World Wide Web

These days, a number of search engines—such as Google, Inc., and Mozilla Firefox—procure information and organize it in a variety of ways to help users navigate the World Wide Web. That is why there are so many different search engines. At a basic level, a search engine is one of two things: a Robot, shown as a letter R, or a Directory, shown as a letter D. Although some search engines combine features of both, most are predominantly R or D.

A Robot uses a software program to search, catalog, and then organize information on the Internet. Organization of information can be completed in a number of ways—among them harvester, robot, spider, wanderer, and worm—and employing diverse ways of searching Websites to gather information.

Directory search engines do not search on the Internet for information but rather obtain it from databases consisting of information placed there by individuals. Because each Directory has its own means of categorizing information, multitudes of them exist.

In March 2005, Google, Inc., released its first official version of its free software for finding information stored on computer hard drives. The software scours hard drives for information contained in Adobe Acrobat's portable document format (known as PDF) as well as in music, video files, and e-mail (In Brief, 2005).

It is interesting to note that on Saturday, May 7, 2005, the Google, Inc., search engine went down for fifteen minutes, from 6:45 until 7:00 P.M. EST. A Google spokesperson said that the problem was not caused by crackers—as many people had thought—but by a problem related to the Domain Name System. He did not elaborate (Google Admin., 2005).

Future Development of the Internet

In 2006, the World Wide Web is being used to:

- Boost electronic business to sell multitudes of product lines, ranging from books and DVDs to expensive jewelry, clothes, and vehicles.
- Market services online, ranging from hotel reservations and airline bookings to conference registrations, wedding arrangements, and executive searching.
- Market oneself online through blog development—an online journal in which the job seekers make reference to their education, work experience, thoughts, and personal quirks that are value-added.
- Produce interactive entertainment activities such as multiuser online gaming.
- Complete financial transactions online rather than visiting a brick-and-mortar facility, or to trade in securities, adjust one's retirement funds, and apply for loans without having to leave home.

The 2005 holiday season indicated the huge success of electronic business. According to recently released statistics, Canadian online retailers posted their best holiday shopping season yet, as more buyers went to the Internet to buy their holiday gifts. The result was hundreds of millions of dollars in sales and another quarter of double-digit growth for Internet shopping portals. The holiday season, it is said, is simply a condensed version of the year's overall online trends and potential. The hot items ranged from high-tech toys like Apple iPods and Microsoft's Xbox 360s to books and gold jewelry. Also, Seattle, Washington–based Amazon.com, Inc., posted a record season. The online retailer said that it got more than 100 million orders during the holiday season, and on December 12, 2005, its busiest day, it processed an average of forty-one items per second. Besides buying gifts for others online, some Internet users sent online holiday cards rather than using traditional cards (El Akkad, 2005).

Also, in the office environment, instant messaging, or IM, is an essential business tool—essentially turbocharged e-mail—that is another good indicator of present-day Internet use. Using

IM, employees can communicate with one another instanta-
neously. From an employer's perspective, IM's benefits include
real-time communications, improved customer service, and
more efficient employee productivity. According to a recent survey by Pew Internet and American Life, today 53 million American adults use IM—and only 11 million of them use it for work
purposes. Most use IM to communicate with friends and family
members. At IBM, all of the company's 320,000 employees use its
Lotus Sametime instant messaging software to communicate
with one another; they apparently send 4 million instant messages each day and hold 12,000 Web conferences each month
with more than 80,000 attendees. In fact, more than 60 percent of
the Fortune 100 companies apparently use IBM's Lotus Sametime software, including most worldwide commercial banks and
automobile makers. America Online, Inc.—which has the largest
instant messaging service, known as AIM—apparently handles 2
billion messages daily. The downside of IM's popularity is that
problems arise because most companies do not have IM policies
in place to manage the rapidly growing tool. Although about
60 percent of all organizations today monitor their employees'
e-mail, only about 10 percent monitor IM (Lorek, 2006).

But is the Web in 2005 all that different from what it was
like, say, five years ago? Most IT people would say that today's
Internet is all about Web 2.0.

No, the World Wide Web is not much different from what it
was five years ago, but it has been evolving. In fact, Web 2.0 today is just more of what it used to be: faster, more interactive,
and more adaptable. High-speed Internet access is more common, a lot of the software has been improved, and there are
many more users online—what some experts are calling the
"network effect" (Ingram, 2005).

So, what is Web 2.0? It is Web-based services—things consumers can use that take advantage of the kind of real-time inter-
activity that software like AJAX (the asynchronous javascript)
and the Extensible Markup Language (XML) allow.

Over the past year, for example, Yahoo.com incorporated
blogs into its search and news functions, and it used the new and
improved AJAX-powered e-mail. Microsoft Corporation has released some early-stage services such as live.com, a new and improved AJAX-powered e-mail client with Web-based antivirus
capacity. And Google's Earth and Google Maps, two applications

using Web 2.0 technology, have helped millions of Internet users to not only get a sense of the devastation caused by Hurricane Katrina but to search for, say, a nearby wine store (ibid.).

Online Voting and VoIP Long-Distance Telephoning

How, then, is the development of the Internet likely to occur?

As noted earlier, online voting is likely to be tested in 2006 in the United Kingdom, and Voice over Internet Protocol (VoIP) technology already permits telephone calls to be placed over networks like the Internet—making it much less expensive to place long-distance calls than when calling with a conventional telephone. The main drawbacks from a convenience point of view are that there is a greater potential for "dropped" calls and a generally reduced voice quality (compared with traditional phones).

Current VoIP products include (Walsh and Kuhn, 2005):

- Telephone handsets: usually more than a simple handset with a dial pad; some of these pieces have a base-station design that provides the convenience of a cordless phone.
- Conferencing units that provide the same type of service as conventional conference-call telephone setups, but because communication occurs over the Net, users can coordinate other communication devices such as a whiteboard displayed on computer monitors at participating sites.
- Wireless mobile units that are becoming quite popular; many organizations already have an installed base of 802.11 networking equipment. However, because of the known vulnerabilities associated with the 802.11b protocols, VoIP equipment can have some major security flaws.
- PC or softphone: using a headset, software, and an inexpensive connection service, users can use any PC or workstation as a VoIP unit or softphone.

In terms of security and trust issues, experts caution that VoIP technology could create an enormous market for voice spam (known as SPIT, or Spam Over Internet Telephony). Using automated VoIP servers, telemarketers could send messages to thousands of Internet telephone addresses all at once rather than placing each call separately using a conventional phone line. Currently, SPIT is not much of a problem, primarily because not enough people receive their telephone services over the Internet. In short, while trade-offs between convenience and security are routine in software, VoIP is not an exception. Just about all VoIP components use integrated Web services for configuration, and while Web interfaces are attractive, easy to use, and inexpensive to produce because of the wide availability of development tools, they pose security risks. Because VoIP device Web applications have weak or no access control as well as script vulnerabilities, there could be privacy and Denial of Service weaknesses. Apparently, there are only about 5 million VoIP users worldwide at this time; time will tell if the strengths of VoIP can outweigh the weaknesses (Avery, 2004; Buckler, 2004; In Brief, 2004).

Beating Jetlag with Redefined Internet Videoconferencing

After the September 11, 2001, terrorist attacks on the World Trade Center, U.S. employees and employers were increasingly reluctant to fly. The same could be said for those at DreamWorks Animation SKG, Inc. Before this incident, somewhere between forty and sixty employees would daily fly up and down the California coast. To deal with the issue, the animation company looked for a high-tech partner who could help solve the problem. Dream-Works and Hewlett-Packard Co. developed an audio-video linkup called "Halo" that makes use of large, fifty-inch plasma TV screens (giving users the sense of sitting on two sides of a window), super rapid fiber-optic connections (reducing delays between sender and receiver), and "smart" software that makes it possible for users to share information on one of the screens, such as Web pages or slides (Avery, 2006).

Halo has the ability to redefine videoconferencing, because unlike older videoconferencing equipment, it can capture the finer points of a face-to-face meeting. The problem is the cost, and for that reason Halo is now being targeted at Fortune 500 companies that can get a decent payback for the costly audio video linkup. The videoconference rooms cost U.S. $550,000 each, and a buyer must have at least two rooms. Also, there is a network and service fee of U.S. $18,000 a month per room (ibid.).

Indulging in the Need for Speed with New and Improved Wireless Networks

Mobile commerce—the conducting of business transactions over Internet-enabled wireless devices—is slowly becoming a dominant force in business and society. The push of advancing technology and the pull of public demand for low-cost, high-speed communications and ubiquitous access to information anytime, anywhere have revolutionized the telecommunication industry over the past twenty years. This revolution has, in fact, led Canada to have one of the lowest Internet access charges among G7 nations. More recently, Internet access and high computing power in wireless devices began to pave the way for the introduction of broadband interactive multimedia applications. Despite these signs of progress, the wireless Web market is still in its infancy; because most Internet users think of wireless networks as slower than wired ones (Grami and Schell, 2004).

But if a Wi-Fi technology standard called 802.11n comes in, that could boost wireless performance to the same level as that found in wired networks. Although the original 802.11b version of Wi-Fi transmits information at up to 11 Mbps, the newer 802.11n standard guarantees speeds of 100 Mbps. The Institute of Electrical and Electronic Engineers (IEEE) has been working on this new and improved standard since 2002, but delays have taken place because of infighting among companies having an interest in the standard. The promised 100 Mbps or greater would bring wireless to an ideal speed for wonderful things like multimedia and large-file transfers. The new standard should be ready for implementation after the year 2007 (Buckler, 2005).

Growing Podcasting

Techies are talking about the technology called podcasting, which makes it easy for Internet users to create their own audio recordings and then post them on the Web. Podcasting allows users to have radio shows about any topic near and dear to their hearts. The more innovative types have experimented with advertisements and subscriptions. In November 2005, pioneer Adam Curry launched a podcast network with thirty to fifty shows that split advertisement revenues. Curry's concept brought him nearly U.S. $10 million worth of investment from two top venture capital firms in Silicon Valley (Green, 2005).

Curry is not alone in his entrepreneurial adventure. Many podcasters are looking for sponsors, typically involving a 15- or 30-second audio advertisement at the start of the podcast. In fact, popular podcasts could see advertisement rates range from a few thousand dollars per month to $45,000 or more. Other podcasters are coming up with alternative financial schemes. "This Week in Tech," which has more than 200,000 listeners, asks each listener for a $2 donation per month. As a result, about $10,000 per month is generated. New developments make podcasting a fascinating new technology on the Internet highway (ibid.).

Turning to the Web to Watch Television

Within five years, consumers who turn on the television to entertain themselves can anticipate alternative ways to watch their favorite shows—such as downloading programs and viewing television streamed over the Internet. There will be television-equipped cell phones, and favorite shows will be pulled from Websites—likely without all the heavy advertising that television viewers face today. Although there has been a large conservative television market, in which viewers watch scheduled shows, and a nontraditional television market of tech-savvy viewers who prefer mobile television and on-demand programming, it is likely that the latter will survive and the former will dwindle. The down side, say marketers, is that without advertisement revenues to cover the cost of producing and distributing local television news, it is likely that significant amounts of programming will move to the Internet and that television news will become available only in major urban areas (Robertson, 2006).

Dual-Core Multi-Processing and 64-Bit Computing

Two or more microprocessors plugged into the same chipset are certainly more powerful than one. That is what dual-core multi-processing is about, and in the future it should become especially powerful in servers and workstations designed for smaller businesses. In fact, by the end of 2007, dual-core will be the microprocessing standard used by Intel Corporation and Advanced Micro Devices, Inc. Dual-core multiprocessing will allow smaller businesses to use applications having an enterprise scale. Small businesses could then do complex applications such as customer relations management, graphics creation and editing, and decision support—all at a cost comparable to that of current, single-core chips.

Moreover, the next generation of computing will arrive in the form of dual-core and 64-bit processors like Intel's Itanium and Xeon chipsets. They will have a huge memory caching—on the scale of 16 billion gigabytes of memory, compared with the 4 gigabytes of memory of the 32-bit systems. The upshot is that with this amount of memory, Internet data searches and data retrieval will be a lot quicker. After Microsoft Corporation launches its Vista operating system, many 64-bit applications are sure to be released onto the market (McLean, 2005).

Conclusion

Without question, a variety of extremely positive services have been made available to citizens around the globe with the development and rapid growth of the Internet over the last decade. These very useful functions range from communications services, such as instant messaging and telephony, to rapid, real-time online transactions, such as e-commerce, Internet-banking, online gaming, political activism, and online voting. Also, within the past few years, physicians have been able to access over the Internet and through handheld wireless devices patients' health histories and diagnostic records without having to rely on time-delaying courier services. Not only have young billionaires been made with the creative development of "Google-like" search engines, but, in addition, governments around the globe have made use of the Internet to collect homeland security intelligence as a means of keeping their citizens safe.

Some important evolutional improvements of the Internet described in the first half of this chapter included the ARPAnet (during the 1960s), electronic mail or e-mail (which first arrived about 1971), remote login (which arrived in 1972), file transfer (from the basic file transfer applications of the 1970s to the sophisticated file-sharing applications of our day), the emergence of the World Wide Web (WWW), the Universal Resource Locator or URL (which arrived sometime between 1989 and 1991), and the birth of the Google search engine. Despite this positive evolution, however, cyber-criminals have used the capacity of the Internet to engage in cyber-crime—and on a much larger scale than had previously been seen.

References

Adler, Joe. "Hackers May Target Pacemaker Technology." InfoSec News, http://lists.jammed.com/ISN/2005/02/0120.html (accessed February 24, 2005).

Arnone, Michael. "Contractors Told to Relax about Blackberry." 101 Communications, http://www.fcw.com/article92023–01–23–06-Print (accessed January 23, 2006).

Avery, Simon. "How to Beat Jet Lag for $550,000." *Globe and Mail*, January 3, 2006, pp. B1, B2.

_____. "New Service Searches for the Right Connection." *Globe and Mail*, November 4, 2004, p. B13.

Berners-Lee, Tim. "The WorldWideWeb Browser." Tim Berners-Lee Website, http://www.w3.org/People/Berners-Lee/WorldWideWeb.html (accessed January 20, 2006).

Brearton, Steve. "Innovation." *Globe and Mail Report on Business* 22 (September 2005): 54–55, 57, 59, 63, 65–66, 68.

Buckler, Grant. "New Standard Would Let Wireless Junkies Indulge Need for Speed." *Globe and Mail*, December 29, 2005, p. B5.

_____. "Spammers on Your Phone Line? Makes You Want to SPIT." *Globe and Mail*, November 4, 2004, p. B15.

Chen, Y-C, et al. "Online Gaming Crime and Security Issue—Cases and Countermeasures from Taiwan." University of New Brunswick, http://dev.hil.unb.ca/Texts/PST/pdf/chen.pdf (accessed October 13–15, 2004).

Computer Industry Almanac, Inc. "Worldwide Internet Users Will Top One Billion in 2005." Computer Industry Almanac, http://www.c-i-a.com/pr0904.htm (accessed September 3, 2004).

Edge. "Networking to Internet, a Long Journey." http://neworder.box.sk/newsread.php?newsid=6047 (accessed September 19, 2002).

El Akkad, Omar. "On-line Stores Set Holiday Records." *Globe and Mail*, December 27, 2005, pp. B1, B8.

Evers, Joris. "Most Spam Still Coming from the U.S." CNET Networks, Inc., http://news.com.com/Most+spam+still+coming+from+the+U.S./2100–1029_3–6030758.html (accessed January 24, 2006).

Farzad, R., and B. Elgin. "Googling for Gold." *Businessweek* 49 (December 5, 2005): 48–52, 54.

Findlaw.com. "U.S. Supreme Court Syllabus: *Reno v. ACLU.*" Findlaw, http://caselaw.lp.findlaw.com/scripts/getcase.pl?court=US&vol=000&invol=96–511 (accessed January 28, 2006).

Gaudin, Sharon. "A Conversation with the Inventor of Email." http://itmanagement.earthweb.com/entdev/article.php/1408411 (accessed July 16, 2002).

Goodwin, Jean. "Child Pornography Prevention Act." Northwestern University, http://faculty-web.at.northwestern.edu/commstud/freespeech/cont/cases/morphed.html (accessed February 6, 2001).

Google Admin. "Google Down? Getting 404! Google Hacked?" Google.com, http://www.submitexpress.com/bbs/post-1601.html&highlight=&sid=cdfcb4b3aa56cdca7df35ed920dd8079 (accessed May 9, 2005).

Grami, A., and B. Schell. "Future Trends in Mobile Commerce: Service Offerings, Technological Advances, and Security Challenges." http://phx.corporate-ir.net/phoenix.zhtml?c=97664&p=irol-news Article&ID=772580&highlight=#top (accessed October 13–15, 2004).

Green, Heather. "Searching for the Pod of Gold." *Businessweek* 46 (November 14, 2005): 95–96.

Haviland, David. "The Transistor." Nobel Foundation, http://nobel prize.org/physics/educational/transistor/history/ (accessed June 20, 2005).

Hewlett Packard Development Company. "Canada's SHARCNET Research Network Chooses HP for Exponential Leap in Supercomputing Power." Hewlett Packard, http://www.hp.com/hpinfo/newsroom/press/2005/050621b.html (accessed June 21, 2005).

ICANN. "Vinton G. Cerf." ICANN, http://www.icann.org/biog/cerf.htm (accessed September 27, 2005).

In Brief. "Google's Official Desktop Search Software Released." *Globe and Mail,* March 10, 2005, p. B10.

———. "Voice Spam Alert." *Globe and Mail,* August 12, 2004, p. B7.

Ingram, Mathew. "Google, Yahoo Go to Top of the Class in Web 2.0 This Year." *Globe and Mail,* December 29, 2005, p. B5.

INRIA. "The French National Institute for Research in Computer Science and Control." INRIA, http://www.inria.fr/ (accessed January 27, 2006).

Lipowicz, Alice. "DHS Vows to Protect Info on National Database." Post-Newsweek Media, Inc., http://www.washingtontechnology.com/news/1_1/daily_news/27812–1.html (accessed January 24, 2006).

Lorek, Laura. "The IM Revolution—and the Risks." *Globe and Mail,* January 25, 2006, p. C11.

McKenna, Barrie. "RIM Wins Key Battle in Patent Dispute." *Globe and Mail,* February 2, 2006, pp. B1, B8.

McLean, Dan. "'Virtualizing' and Other Trends to Come in IT World." *Globe and Mail,* December 22, 2005, p. B6.

MercuryNews.com. "Hacker Pleads Guilty to Building, Renting Attack Network." Mercury News, http://www.mercurynews.com/mld/mercurynews/business/13693354.htm (accessed January 23, 2006).

Newkerala.com. "22 Police Officers Complete Anti-Terror Training." Newkerala.com, http://www.newkerala.com/news.php?action=fullnews&id=93688 (accessed January 24, 2006).

Palmer, Ian. "Workplace: It's Not Just the Big Boys Using Intranets any Longer." *Globe and Mail,* May 5, 2005, p. B27.

Phrackstaff@phrack.org. "Call for Papers—*Phrack* #63." *Phrack,* http://phrack.org (accessed January 24, 2005).

Reti, Kalman. "Symbolics." Symbolics, http://www.symbolics.com (accessed January 19, 2006).

Robertson, Grant. "Internet Threatens Canadian Broadcasters: Report." *Globe and Mail,* January 27, 2006, p. B3.

SharpenedNet.com. "Glossary: HTTP." SharpenedNet.com, http://www.sharpened.net/glossary/definition.php?http (accessed January 25, 2006).

Stone, Tim. "AMZN Investor Relations." Internet Stock Blog, http://internetstockblog.com/article/6471 (accessed February 2, 2006).

Storer, T., and I. Duncan. "Practical Remote Electronic Elections for the U.K." http://www.dcs.st-and.ac.uk/research/publications/SD04c.php (accessed October 6, 2005).

Sun Microsystems, Inc. "The Fab Four Reunites." Sun Microsystems, http://www.sun.com/2006–0113/feature/index.html (accessed January 19, 2006).

Vaagan, R., and W. Koehler. "Intellectual Property Rights vs. Public Access Rights: Ethical Aspects of the DeCSS Decryption Program." *Information Research* 10 (April 2005): 1–10.

Walsh, T., and R. Kuhn. "Challenges in Securing Voice over IP." *IEEE Security & Privacy* 3 (May/June 2005): 44–49.

W3C. "Extensible Markup Language (XML) 1.0." W3C, http://www.xml.com/axml/testaxml.htm (accessed February 10, 1998).

2

Problems, Controversies, and Solutions

With the growth of and diversity in Internet traffic, a dark side to the Internet has surfaced in recent decades, particularly since the late 1980s. The first half of this chapter explores a number of privacy, security, and trust issues of concern to various segments of society. Some technical defenses are discussed. A number of crimes perpetrated by abusing the Internet infrastructure are described, such as illegal file transfer (for example, of copyrighted material and intellectual property), transfer of censured material (such as child pornography), identity theft, and fraud (such as through online auction abuses). This chapter also explains various laws passed in the United States and elsewhere to reduce such abuses.

Internet Users' Concerns about Privacy, Security, and Trust

In the last decade in particular, the growth of the Internet and e-commerce has dramatically increased the amount of personal information that can potentially be collected about individuals by corporations and governments. Such data collection, along with usage tracking and the sharing of data with third parties—especially in light of the fact that such actions can easily be done through high-speed links and high-capacity storage devices without the consumer's expressed knowledge or consent—has raised issues among Internet users about privacy, security, and trust (Grami and Schell, 2004).

Concerns about Privacy

Internet Usage Policies.   Before giving employees access to the Internet, companies, government agencies, medical institutions, and universities typically have would-be Internet users sign a required "Internet Usage Policy," to make users accountable for their online activities and to prevent abuses and liability claims.

Such a form typically looks like this:

> I have received a copy of Company X's Internet Acceptable Use Policy. I understand this policy's terms and conditions and agree to follow them. I understand that Company X's software may record for management's review the Internet addresses of all the Websites I visit. I also understand that management may maintain a record of all of my network activity (including the sending and receiving of e-files).
>
> I acknowledge that all e-files and e-messages sent or received by me may be recorded and stored in an archive file for management's review. I fully understand that if I violate this policy, I can receive disciplinary action, ranging from the revoking of my Internet privileges to firing. If I violate this policy in a criminal way, I understand that I may also face criminal charges.
>
> Employee Signature _____ Date: _____
>
> Employee Name (Print) _____

While, on face, the form seems to be reasonable from a managerial and institutional liability point of view, it has raised concerns among the signers that their privacy may be at risk.

Privacy Defined.   Privacy, by definition, is the state of being free from unauthorized access. By signing the form, the Internet users are agreeing to let management have authorized access to all of their network activities, for the primary reason of keeping the company safe from liability claims. At the same time, Internet users are assuming in good faith that management will not be sharing their online activity records with third parties who have no need to know.

Privacy laws, from a macro level, deal with the right of individual privacy, critical to maintaining the quality of life that citizens in a free society expect. Privacy laws generally maintain that an individual's privacy shall not be violated unless the government (or a company) can show some compelling reason to do so—such as by providing evidence that the safety of the nation (or that of a company) is at risk. This tenet forms the basis of privacy laws in the United States and elsewhere. Privacy also means being able to maintain a balance between individuals' privacy rights and those of the government in providing national security or the rights of a company in terms of avoiding liability claims. From a citizen's point of view, privacy is the interest an individual has in controlling, or at least in significantly influencing, the handling of personal data (Al-Fedaghi, 2005).

Privacy Policies.   To balance employees' individual privacy rights and those of the company, management typically prepares "privacy policies" for their organizations. A privacy policy, by definition, is a clear description of how companies use e-mail addresses and other information they gather when online users opt to be included in requests for company information, newsletters, or third-party deals. U.S. state laws compel companies not only to state their privacy policy on their Websites but also to place the policy statement so that people may plainly see it. In some cases, state laws may also prescribe the display form for the policy.

Internet Privacy Issues.   From a technical perspective, privacy issues in the security sense include digital rights management, spam deterrence, anonymity maintenance, and cracker disclosure rule adequacy. It is interesting to note that over the past few years, intruders into companies' and institutions' computer networks have violated the privacy rights of employees and online registrants. As noted in Chapter 1, in 2005 evidence surfaced of the poorly regulated Internet information brokerage industry, for on or about March 10, cyber-criminals stole passwords from legitimate online users of as many as 32,000 Americans in a database owned by the renowned LexisNexis Group.

Similar computer network breaches have also occurred at ChoicePoint, Inc., and at the Bank of America. These intrusions prompted calls for U.S. federal government oversight through the General Services Administration (GSA) to look into the matter. The GSA agreed to review the online security policies of the

Bank of America and four other SmartPay contractors providing credit avenues for online transactions—Bank One of Delaware, Citibank of New York, Mellon Bank of Pittsburgh, and the U.S. Bank of Minneapolis. Recommendations were to follow about providing adequate protection for the safety of federal employees' personal information. Fears of identity theft surfaced on a large scale.

The present-day reality is that regardless of how well intentioned management is about protecting employees' privacy rights, valuable personal information can often be collected by hidden online tools such as cookies and Web bugs—and then be shared with third parties for marketing purposes or surveillance. Contrary to what some individuals think, cookies are not by themselves a security risk. They are simply small bits of data—a message—transmitted from a Web server to a Web browser. The browser stores the message in a text file, and each time the browser requests from the server a particular page, the message is sent back to the server. Cookies personalize a Website for users. That is, when users enter a Website, they may be asked to complete forms indicating their name and certain particulars. Instead of seeing a generic welcome page, users are later greeted with a page including their identifiers stored in the cookies.

Nevertheless, as noted, there is controversy surrounding cookies. For example, cookies may be accessed, read, and used by malicious Websites unintentionally visited by innocent users. This cookie information can be used to gather intelligence on the user and later be utilized against the user, it might be combined with other off-line data, such as demographic and psychographic information, to predict a user's interests, needs, and possible future purchases.

Proposed Solutions.   How might these privacy concerns be resolved? That is an especially important question in light of the fact that public surveys have indicated that many consumers are still afraid to buy things online because they fear that their credit card numbers will be used by someone else, even though the credit card companies say that they will not hold consumers accountable for fraudulent charges. Trust seals and increased government regulation are two ways of promoting improved privacy disclosures on the Internet.

We've all seen trust seals on e-business Websites—green Truste images, the BBBOnLine (Better Business Bureau OnLine) padlocks, and a host of other privacy and security seals. In fact, more than 2,000 companies are paying as much as $13,000 per year to display those logos on their Websites. Almost half of the Fortune 100 companies display such logos, and of the fourteen IT sites in the Fortune 100 companies, ten have such seals (Cline, 2003). In terms of market share, Truste (now nine years in business) lists 1,374 Websites, while BBBOnLine lists 701 Websites (ibid.).

But do they really work? Although trust seals are intended to advance privacy for consumers through legislation—primarily through self-regulation by businesses—critics say that they can become more of a privacy advocate for corporations than for consumers. Both Truste and BBBOnLine charge companies annual fees based on yearly revenues. With Truste, companies having revenues of less than $5 million are charged about $600, while those having revenues of more than $2 billion pay about $13,000. The charge is lower with BBBOnLine; smaller businesses pay only about $200, while companies with revenues over $2 billion pay about $7,000. Companies having trust seals say that they will follow the trust standards, they will provide online clients with a way to opt out of direct marketing and to having their personal information sold to third parties, and they will be given a means to access the company's information and file complaints. ScanAlert, an emerging security seal provider, attributes Internet sales increases of 10 to 33 percent to having that seal on a company's Website (ibid.).

From a consumer perspective, a recent study conducted by Flinn and Lumsden (2005) indicated that 42 percent of the consumers in their study reported that they were more likely to trust a Website that displays a trust mark than one that did not have the display, and 49 percent of the consumers said that they are likely to trust a Website only if they are able to recognize the trust mark program.

Moreover, while government regulations are increasing to advance the privacy of citizens—as through the passage of the Canadian Personal Information Protection and Electronic Documents Act (PIPEDA) and the U.S. Health Insurance Portability and Accountability Act of 1996, citizens themselves are often uneasy with these measures, for they dislike having, say, their telecommunications traffic monitored by government agents.

As a case in point, the U.S. Carnivore DCS 1000 program (a computer-automated snooping tool capable of intercepting and sorting through millions of messages like telephone calls and e-mails passing through ISPs by monitoring incoming and outgoing messages to specific IP addresses, particularly those of suspected terrorists) was never implemented by the U.S. government. Why? There was massive public outcry that citizens' privacy rights would be violated with such an implementation. Instead, the Federal Bureau of Investigation (FBI) moved to using commercially available software and encouraged Internet providers to conduct wiretaps on suspected individuals and to forward their findings to government agents.

As a result of privacy concerns, the field of Information Ethics (IE) has developed, dealing with issues arising from connecting technology with concepts such as privacy, Intellectual Property Rights (IPR) information access, and intellectual freedom. Although IE issues were raised as early as 1980, the field has evolved into a multithreaded phenomenon, stimulated by the concerns of a number of disciplines over abuses occurring through the Internet. According to IE, information itself, in some form or role, is recognized to have intrinsic moral value.

Theoreticians have formulated a number of complex mathematical solutions for providing better information protection over the Internet. One of these is provided in the database area and is based on atomic assertions, such that the relationship between individuals and their own atomic private information is identified through the notion of proprietorship. The concept goes like this. Suppose that a company has a private information database that includes information privacy databases for three departments and information privacy databases for each of the company's employees. Each possession of a department contains the private information about its employees and perhaps employees of other departments. It is conceptually possible to enforce global privacy constraints on information in possession (Al-Fedaghi, 2005).

Also, all employees have an information privacy database called Proprietary.Known that tells them what, for example, department 1 "knows" about them. Because the database automatically inserts any private information in the appropriate employee's Proprietary.Known, this capability provides some verification that the employee is aware of the personal information held by the department. The same verification process would hold for the other two departments (Al-Fedaghi, 2005).

Concerns about Security

Wireless devices give ready access to the Internet for owners with cellular phones and handheld computers, and for that convenience factor alone, mobile commerce services are appealing to consumers around the world. However, the success of mobile commerce depends on the security of the underlying mobile technologies. Without such security, citizens have concerns about using these devices.

Without getting too technical, wireless communications rely on open and public transmission media (over the air) that raise security vulnerabilities. These vulnerabilities for wireless devices, in particular, must be addressed by the companies releasing such products into the marketplace, in addition to providing fixes for software vulnerabilities found in wired networks.

The mobile commerce challenges relate to the user's mobile device, the wireless access network, the wired-line backbone network, and mobile commerce software applications. Unlike wireline networks, the unique traits of wireless networks pose a number of complex challenges for security experts, such as vulnerability of the air interface, an open Peer-to-Peer (P2P) network architecture (in mobile ad hoc networks), a shared wireless medium, the limited computing power of mobile devices, a highly dynamic network topology, and the low data rates and frequent "disconnects" of wireless communications. Coping with these vulnerabilities costs companies money; for example, the chargeback rate for credit card transactions is about fifteen times higher than the chargeback for instore point-of-sale credit card transactions (Grami and Schell, 2004).

Security Defined.   "Security" means being protected from one's adversaries, particularly from those who would do harm—even unintentionally—to property or to a person.

Information Technology security issues, in particular, include but are not limited to adequate authentication of users while online, adequate critical infrastructure protection, well-designed disaster recovery plans following a cyber exploit, sound network intrusion detection and network management, adequate and regularly updated malicious code software protection, adequate physical security protection of networks, the development and implementation of sound security policies, and state-of-the-art wireless security protection.

Internet Security Issues.   Security breaches in the computer networks of governments, businesses, financial institutions, and educational institutions occur daily, with some of them making headlines and with some of them shaking consumers' and shareholders' confidence.

On January 30, 2005, for example, an Internet-related security incident brought considerable embarrassment to the Dutch armed forces. About seventy-five pages of highly classified documents about human traffickers from the computers of the Dutch Royal Marechaussee (the armed forces contingency that guards the Dutch borders) somehow found their way to the controversial Weblog (online log) of Geen Stijl (meaning "No Style").

The conjecture is that a Dutch armed forces staff person worked on the documents at home and unwittingly shared the contents of his computer's hard drive with numerous others when he logged onto Kazaa—a nonsecure online file-sharing Website. And that was not the first time that the Dutch government had made news media headlines over computer security issues. In 2004 the Dutch public prosecutor's office was equally embarrassed after it was publicized that the prosecutor had thrown his old PC into the trash, making available for public scrutiny his hard drive with hundreds of pages of classified data on high-profile Dutch crimes as well as his own credit card numbers and personal tax information. As a result, the prosecutor resigned from his job (Libbenga, 2005).

Mobile services, in general, are prone to two types of fraud risks as a result of security vulnerabilities: subscription fraud and device fraud. Subscription fraud (more commonly known as "identity theft") is the same problem that issuers of credit cards have when someone pretends to be another subscriber. As with other types of credit-related identity theft, an impostor fails to pay the bills and the service is eventually stopped. Although the customer will not be responsible for paying an impostor's bills, credit rating complications can arise (Grami and Schell, 2004).

Device theft has become more attractive to thieves as wireless devices become smaller but more powerful. Today, devices have location technology embedded within to assist authorities in tracking the stolen item. Also, to combat theft, besides the usage password, devices are tailored to owners using effective but inexpensive biometric control technologies (ibid.).

Security Checklist. Given these security concerns of consumers, a security policy checklist detailing the kinds of measures companies generally take to ensure that their networks are secure (and posing key questions for system administrators to answer) is given below. These measures include but are not limited to the following (Queeg Company, 1997):

- *Administrator rights and responsibilities:* Under what conditions may a system administrator examine an employee's account or e-mail, and what parts of the system should the system administrator not examine (for example, looking at bookmarks)? Can the system administrator monitor network traffic, and, if so, what boundaries exist?
- *Back-ups:* What systems in this company are backed up, and, if so, how often? How are back-ups in this company secured and verified?
- *Connections to and from the Internet:* What computers should be seen from the outside? If computers are outside the firewall (bastion hosts), how securely are they separated from computers on the inside? Are connections from the Internet to the internal network allowed, and, if so, how are they authenticated and encrypted? What online traffic is allowed to go outside the internal network? If there is traffic across the Internet, how is it secured, and what protection is in place against worms, viruses, or hostile java applets?
- *Dial-up connections:* Are dial-up connections allowed, and, if so, how are they authenticated and what access level to the internal network do dial-in connections provide? How are modems distributed in this company, and can employees set up modem connections to their home or desktop computers?
- *Documentation:* Does a map of the network topology exist, and is it clearly stated where each computer fits on that map? Is there an inventory of all hardware and software in this company, and does a document exist detailing the preferred security configuration of every system?
- *Emergency procedures:* What kinds of procedures exist for installing security patches or for handling exploits? In cases of system intrusion, is it company policy to shut

down the network immediately, or does the company prefer to monitor the intruder for a time? How and when are employees notified of network intrusions, and at what stage are law enforcement agencies called in?

- *Logs:* What information in this company is logged, and how and where is the information logged? Are the information logs secure from tampering, and, if so, are they regularly examined, and by whom?
- *Physical security:* Are systems physically protected from outsider crackers and adequately secured, where needed, from insider crackers? Are reusable passwords utilized internally, and are employees told through company policy to change their passwords routinely?
- *Sensitive information:* How is sensitive and proprietary information protected online, and how are backup tapes protected?
- *User rights and responsibilities:* How much freedom do employees have in terms of selecting their own operating system, software, and games for their computers, and can employees in this company send and receive personal e-mail or do personal work on company computers? What policies exist regarding resource consumption (for example, disk or CPU quotas) and abuse (accidental or intentional) of services? What penalties exist, for example, if an employee brings down a server?

Proposed Solutions. Security threats in mobile commerce may be passive, such as information monitoring and release for fraudulent purposes, or active, such as the modification of information through Denial of Service (DoS) and unauthorized access. Additional security issues posed by wireless networks have been outlined above.

Generally, businesses and government agencies take two kinds of approaches to preventing security breaches: proactive approaches, such as preventing crackers from launching attacks in the first place (typically through various cryptographic techniques); and reactive approaches, which means detecting security threats after the fact and applying appropriate fixes. Comprehensive network security solutions tend to integrate both approaches. In technical circles, securing Websites generally refers to the use of SSL/TLS technology for encrypting and authenticating HTTP connections (Flinn and Lumsden, 2005).

Furthermore, because network security is a chain, it is only as secure as its weakest link. Although enhanced network security features are desirable, they cost money—which is why some companies and institutions, especially the smaller ones, are reluctant to apply an enhanced security solution. Some costs are associated with additional overhead (such as increased bandwidth), increased complexity (which requires specialized security experts), and information processing delays (degraded performance), which can, in turn, degrade network performance (Grami and Schell, 2004).

A number of commercial tools have been developed to assist network administrators in preventing intrusions. One such tool was designed in 1995 by Dan Farmer and Wieste Venema and is known as Security Administrator Tool for Analyzing Networks, or "SATAN." SATAN, though a UNIX-based tool, was first designed for SunOS/Solaris and Irix. Ports to many other varieties of UNIX now exist, including one for Linux—permitting any individual with a personal computer and a Slip/PPP account to get information provided by SATAN. This tool works by procuring as much data as possible about system and network services, and it procures data on known software glitches, network configurations, and poorly set up network utilities. Upon discovering vulnerabilities, SATAN gives rather limited data about fixing the problem, but acknowledging that limitation, it is a useful tool for testing single computers or entire networks. Its successor, known as SAINT, is also on the market (Center for Education and Research in Informance Assurance and Security, 1995).

Also, some commercial products, such as Microsoft's Internet Explorer browser, divide the Internet into security zones so that users can assign a Website to zones having suitable security levels. In fact, users can ascertain a Web page's zone by viewing the right side of the browser's status bar. When a user tries to download information from any Website, Internet Explorer reviews the security configuration for that Website's zone. Here are the four zones (Prescription Pricing Authority, 2006):

- *Local Intranet* zone. This zone has addresses not requiring a proxy server, and the addresses here are configured by the system administrator in the Internet Explorer Administrator's Kit (IEAK). By default, the security level of this zone is Medium.

- *Trusted Website* zone. This zone has Websites that users should be able to trust, meaning that they should be able to download or run files without worrying about damage being caused to their computer or information. Users can assign Websites to this zone, whose default security level is Low.
- *Restricted Website* zone. This zone has Websites that users would not trust, because they can not be sure that they can download or run files without damaging their computers or information. Although users can assign Websites to this zone, it defaults to the High security level.
- *Internet* zone. This zone has information not on the user's computer, not on an Intranet, and not assigned to any other zone. This level's default security level is Medium.

From a standard perspective, the emerging IEEE 802.11i standards will improve wireless security issues, in particular, and turn wireless networking into a trusted medium for all users, including preventing Denial of Service problems, caused when the entire network is jammed. The jamming attack could be against the client's wireless device or against the network's access point. Jamming has been difficult to stop, because most wireless local area networking technologies use unlicensed frequencies and are subject to interference from a variety of sources. To prevent unintentional jamming, site surveys are recommended; to prevent intentional Denial of Service (DoS) attacks by crackers, jamming equipment must be identified and removed (Grami and Schell, 2004).

Finally, countries around the world have passed legislation to prevent cyber security breaches and to impose stiff penalties for crackers. One such act was introduced in 1986 in the United States, called the Computer Fraud and Abuse Act. On a global scale, in 2001, group action was taken to combat security breaches. On November 23, the Council of Europe opened to sign its newly drafted Council on Cybercrime. The convention was signed by thirty-three states after the council recognized that many cyber-crimes could not be prosecuted by existing laws—which were typically local in jurisdiction. The convention was, indeed, the first global legislative attempt to set standards on the definition of cyber-crime and to develop policies and procedures to govern international cooperation to combat cyber-crime on the Internet.

Concerns about Trust

The essence of a business transaction conducted either through a wired or a wireless network is trust, and trust must be mutual. Trust in a business context may be expressed in laws, contracts, regulations, policies, and in personal reputations and long-term relationships. Transferring trust to an online environment is not easy. In fact, evidence indicates that many people using the Internet are too trusting of persons connected to the Net. It is not uncommon for Internet users to download virus-infected software or to engage in online chat rooms with strangers. Oddly, though, consumers seem to be wary of trusting online businesses with transactions conducted over the Net.

The Definition of Trust. Trust, a complex concept studied by scholars from a number of academic disciplines, is present in a business relationship when one partner willingly depends on an exchanging partner in whom one has confidence. The term *depend* can take on a number of meanings in this context, including the willingness of one partner to be vulnerable to the actions of the other partner, or the expectation of one partner to receive ethically bound behaviors from the other partner. Security issues regarding Information Technology center on maintaining trust in e-commerce transactions.

Internet Trust Issues. Marketers who use the Internet to "spam" users' e-mail accounts with unwanted advertising using false but legitimate-looking headers have produced countless headaches for online users and have placed serious concerns in consumers' minds about the trustworthiness of the Net. Many efforts have been made to eradicate spam—including the production of filters to stop it from getting through the network and the passage of laws—but, to date, none have been successful.

The root cause of spam is the same property that makes e-mail so useful: the low cost of communicating with a large number of people around the globe. Moreover, the almost zero cost of creating and spoofing an e-mail identity ensures that even when the sending of unwanted bulk messages is prohibited by law—such as the U.S. CAN-SPAM Act of 2003—or an ISP's policies, tracing and punishing the offender is not an easy task.

The reason for the difficulty? Primarily, it is that the foundation of current e-mail systems was not designed with secure authentication in mind. Proposed solutions attempting to remedy this problem have been dismissed as unfeasible in the short term (Seigneur et al., 2004).

An interesting case of breach of online trust occurred in March 2005. The Harvard Business School administration said that, as a result of unauthorized intrusions, they were going to reject 119 applicants who had followed a cracker's instructions to break into the school's admission Internet site to see if they had been accepted into the university. The actions were cited by the school's administration not only as being unethical but also as breaching trust. Other universities have taken similar punitive approaches to such breaches, including Carnegie Mellon University's School of Business. These universities and others similarly affected used the ApplyYourself online application and notification software (Associated Press, 2005).

As noted, a major barrier to the success of online business has been the fundamental lack of faith between business and consumer partners. This lack of trust by consumers is largely caused by their having to provide detailed personal and confidential information on request. Moreover, consumers fear that their credit card number could be used for purposes other than the one allowed.

But concerns about the consumer are also present from the business partner's vantage point. For instance, the company is not sure if the credit card number the consumer gives is genuine, is in good credit standing, and actually belongs to the consumer trying to make the transaction.

In the modern information society, large open and distributed networks—which is what the World Wide Web is—leave Internet applications such as electronic mailing, Peer-to-Peer file sharing, Internet phoning with Voice over Internet Protocol, online auctions like eBay, and online gaming vulnerable to trust issues. From the perspective of a single user, a common security problem of such networks is that most or all other users on the network are unknown. Therefore, communicating with unknowns through the Internet elicits two crucial sets of questions that the user must reflect upon and answer: What is the real identity of other persons on the Net, and can their identities be authenticated? How reliable are other persons on the Net, and is it safe to interact with them? (Jonczy and Haenni, 2005).

Proposed Solutions. The question of identity and authentication deals with the authenticity of the available information regarding other Internet users' identities, while the second deals with whether the service provider is trustworthy—a trust management issue.

In short, authentication is the process of identifying an individual, message, file, and other data. The two major roles for authentication, therefore, are: (1) confirming that users are who they say they are, and (2) confirming that the transmitted Internet message is authentic and has not been altered or forged.

In recent years, a number of products have been developed to assist in the authentication process, including the following:

- *biometrics* (assessing users' signatures, facial features, and other biological identifiers);
- *smart cards* (having microprocessor chips that run cryptographic algorithms and store a private key);
- *digital certificates* containing public or private keys—the value needed to encrypt or decrypt a message; and
- *SecureID,* a commercial product using a key and the current time to generate a random number stream that is verifiable by a server—thus ensuring that a potential user puts in a verifiable number on the card within a set amount of time (typically 5 or 10 seconds).

Trusted authentication management in a distributed network like the Internet—pertinent to the second question raised—is not easy. Trust management is generally based on either a centralized or a decentralized model. In the former, the responsibility of issuing various types of credentials—digitally assigned statements or attestations about what another user is or does—can be taken over by a central authority. Then, all network users place their trust in the central authority (ibid.).

In a decentralized authority, in contrast, every user is also a potential issuer of credentials, and a given set of credentials, perhaps issued by many different users, forms a credential network. Therefore, a web of trust model and solution was introduced by Pretty Good Privacy (PGP), a popular application for e-mail authentication. In short, PGP organizes public keys and corresponding certificates in local key rings. The owner of the key ring gets a web of trust by assigning trust values to all certificate issuers. This web of trust acts as the basis for a qualitative evaluation of the authenticity of the public keys involved. In PGP, the evaluation of a web of trust is founded on three rules and the production of two outcomes: valid or invalid (ibid.).

Unfortunately, to date, authentication systems like PGP and S/MIME, designed to run over the legacy system, have failed to gain large acceptance and to solve real-world trust problems such as spam, because they suffer from a number of deployment usability issues as well as trust management issues. For example, in "web of trust" style systems, Internet users must validate keys out-of-band—a laborious task. Furthermore, while Certificate Authority (CA) schemes replace the onerous task for individual users to check identities, the charges imposed by the CA often act as a barrier to adoption (Seigneur et al., 2004).

As noted, security experts have found that wireless communications, in general, are not as reliable—and, therefore, not as trustworthy—as wired communications. Thus the occurrence of a technical or technological failure is more likely for mobile commerce (conducted with a wireless device) than for electronic commerce (conducted with a wired device).

Technological failures further diminish the level of trust that consumers have in wireless devices. For example, dropped calls (whereby a carrier fails to hand off a call in progress), busy signals (caused by having too many customers in a cell call at the same time), and dead spots (an area where the signal between the handset and the cell tower is blocked) can all degrade the wireless service's performance—and thereby reduce the consumers' trust in the service (Grami and Schell, 2004).

Emerging advances in mobile commerce—whether they are through improved telecommunications technologies to help realize higher rates, wider coverage, and higher quality of service, or through an improved business framework—cultivating measures such as informed consent, minimum risk insurance, improved Website quality and trust, information clarity, company competence and integrity, and public and private policies all help to build consumers' trust in mobile commerce (ibid.).

It is important to note that in recent years there has been some effort by experts to set standards and indicators for a more systematic and coordinated fashion to capture the trustworthiness state of a particular Information Technology infrastructure. The World Wide Web is no exception. Such indicators would reflect the "assurance" of the IT infrastructure to reliably transfer information (including security, quality of service, and availability of service)—thus increasing trust in the network by consumers.

These indicators could then be used to identify areas of the infrastructure requiring attention and be used by an IT organization to assess the return on investment for improved IT infrastructure equipment purchase. Despite the existing work that is in progress, there is currently no standard or widely accepted method of assessing assurance levels associated with IT infrastructures, including end-hosts, servers, applications, routers, firewalls, and the network permitting the subsystems to communicate. Clearly, this is an area where academics and security experts need to focus (Seddigh et al., 2004).

An Overview of Internet Crimes

Crimes committed over the Internet fall in the general category of "cyber-crimes," because they are committed in cyberspace and involve breaches of privacy, security, and trust. Harm resulting over the Net can be either to persons or to property. There are also technical nonoffenses that are politically motivated and that no legislation declares as unlawful (Brenner, 2001).

Internet Crimes Resulting in Harm to Property

Internet cyber-crime resulting in property harm is generally accomplished with well-honed cracking skill sets and includes common variations like the following (Schell and Martin, 2004):

- *Flooding*—a type of Internet vandalism resulting in Denial of Service to authorized, legitimate users of a Website or computer network.
- *Virus and worm production and release*—a type of Internet vandalism causing corruption, and possibly erasing, of data stored on the network.
- *Spoofing*—a type of appropriation of an online user's identity by others online, causing fraud or attempted fraud in some cases, as well as critical infrastructure breakdowns in other cases.

- *Phreaking*—a type of Internet fraud consisting of using technology to make free telephone calls.
- *Infringing Intellectual Property Rights and Copyright*—a type of Internet theft involving the copying of the information or software of others without their consent.

Internet Crimes Resulting in Harm to Persons

Internet cyber-crime resulting in personal harm includes two major variations (ibid.):

- *Cyber-stalking*—using the Internet to try to control, harass, or terrorize another online user to the point that the stalked person fears harm to reputation or to person, either to self or to others.
- *Cyber-pornography*—using the Internet to possess, create, import, display, publish, or distribute pornography (especially child pornography) or other obscene materials.

Technical Nonoffenses

Politically motivated, controversial online behaviors that are technical nonoffenses by law include the following (ibid.):

- *Hacktivism*—using the Internet to promote political platforms or missions.
- *Cyber-vigilantism*—using the Internet to conduct vigilante activities.

It should be noted that cyber-terrorism and terrorism were also in the nonoffense category before the U.S. Congress hastily passed the antiterrorist USA Patriot Act in 2001. The act was passed within seven weeks of the September 11 attack on the World Trade Center. Formerly, terrorists who caused harm to persons or property were charged under other applicable laws, such as homicide, assault, and property destruction.

The Patriot Act, approved overwhelmingly by Congress shortly after the attacks, has long been the subject of fierce debate over the balance between civil liberties and national security. Besides other measures, the act expanded government search and surveillance authority. Just before breaking for Christmas in December 2005, the U.S. House and Senate approved a five-week extension of the Patriot Act, until February 3, 2006. The act had been scheduled to expire on December 31, 2005. On February 1, 2006, the House of Representatives prepared to extend the antiterrorism law to March 10, 2006, giving House and Senate negotiators another five weeks to dispute the controversial statute. The act was signed into law on March 9, 2006.

It should also be noted that on January 10, 2003, after the passage of the Patriot Act, Attorney General John Ashcroft sent to some of his government colleagues a draft of the Patriot II Act, also known as the Domestic Security Enhancement Act of 2003, which would have had more than 100 new provisions intensifying measures against cyber-terrorists. Once word of this controversial act hit the press, public outcry killed its passage. Citizens' concerns can be found at this Website: http://www.eff.org/Censorship/Terrorism_militias/patriot-act-II-analysis.php.

Internet Criminal Liability: Four Elements

Conventional Crimes and the Four Elements. For old-fashioned/conventional or Internet crimes to occur, Anglo-American law bases criminal liability on the coincidence of four elements (Brenner, 2001):

- A culpable mental state (the *mens rea*).
- A criminal action or a failure to act when one is under a duty to do so (the *actus reus*).
- The existence of certain necessary conditions, or "attendant circumstances." With some crimes, it must be proven that certain events occurred, or certain facts are true, in order for a person to be found guilty of a crime.
- A prohibited result, or harm to property or to person.

The conventional, real-world crime of bigamy, for example, illustrates nicely how all of these elements must combine for the imposition of liability. To commit bigamy, an individual must enter into a marriage knowing either that he or she is already married, or that the person whom he or she is marrying is already married. The prohibited act, then, is the redundant marriage (the *actus reus*). The culpable mental state (the *mens rea*) is the perpetrator's knowledge of entering into a redundant marriage. The attendant circumstance is the existence of the previous marriage still being in force. Finally, the prohibited result, or harm to another person, is the threat that bigamous marriages pose to the stability of family life.

Simply put, a conventional crime, as well as an Internet crime, involves conduct that is unacceptable to society's standards. Therefore, society through its laws, imposes criminal liability. In 2001, Brenner said that except for bigamy and sexual assault—which technically cannot be committed through the Internet, because they are truly real-world acts—other conventional crimes seem to be able to make a smooth transition into the virtual world.

This does not suggest, however, that there has been an absence of controversy around virtual assault cases. For example, Donn Parker debated Brenner's position that bigamy should not be allowed to take place through the Internet. He said that it may, in fact, soon become possible for the conventional crime of bigamy to be perpetrated through the Internet. As digital signatures become valid, a marriage ceremony could be conducted with the two parties and an official in three different places. Marriage using the Web would make for a great reality television show—and allegations of fraudulent marriages could soon follow (Parker, 2001).

Some Differences between Conventional Crimes and Internet Crimes.
At times, there can be some differences between real-world, conventional crimes (even if they are relatively new) and those conducted through the Internet. Using stalking and cyber-stalking as a new crime case in point, there are some striking legal similarities between the two, and some striking differences.

Although online harassment and threats can take many forms, cyber-stalking shares important characteristics with offline stalking. These common traits include the following (U.S. Department of Justice, 2003):

- First, many stalkers are motivated by a desire to exert control over their targets and engage in similar types of behavior to accomplish this end.
- Second, as with offline stalking, the available evidence suggests that the majority of cyber-stalkers are men, and the majority of their targets are women, although there have been cases reported of women cyber-stalking men and of same-sex cyber-stalking.

- Third, often the cyber-stalker and the target had a prior relationship, and the cyber-stalking begins when the victim attempts to break off the relationship. However, there also have been many instances of cyber-stalking by strangers.

Given the enormous amount of personal information available through the Internet, a cyber-stalker can easily locate private information about a potential target with a few mouse clicks.

The fact that cyber-stalking does not involve physical contact may create the wrong impression that it is a more benign act than physical stalking. That is not necessarily true. As the Internet becomes an ever more integral part of our personal and professional lives, stalkers can take advantage of the ease of communications as well as increased access to personal information. In addition, the ease of use and nonconfrontational, impersonal, and sometimes anonymous nature of Internet communications seems to remove disincentives to cyber-stalking. Whereas potential stalkers in real life may be unwilling or unable to confront a target in person, they may have little hesitation in sending threatening e-mails to the target. Finally, as with physical stalking, online harassment and threats may be a forerunner to more serious behavior, including physical violence (Schell and Lanteigne, 2000).

Differences between conventional stalking and cyber-stalking, however, have been noted. These include the following (U.S. Department of Justice, 2003):

- First, offline stalking generally requires the perpetrator and the target to be located in the same geographical area, whereas cyber-stalkers may be next door or around the world.
- Second, the Internet makes it easier for a cyber-stalker to encourage third parties to harass or threaten a target by impersonating the target online and then posting inflammatory messages in chat rooms, encouraging online viewers of that message to send threatening messages back to the targeted "author." Alternatively, the messages sent could encourage viewers to perform harm (such as sexual assault) to the target, who would be painted as consenting to the harmful act.

- Third, the Internet seems to lower barriers to harassment and threats, for cyber-stalkers do not need to confront the victim physically; they could encourage others to complete such acts.

Special stalking laws have been passed in recent years to cope with this "new" crime. The first traditional stalking law was enacted by the state of California in 1990, and since then, Canada, the United Kingdom, and Australia have passed similar legislation. A number of U.S. states have put in special statutes for dealing with cyber-stalking relating to the Internet and electronic transmission of threatening communications, including Alabama, Arizona, Connecticut, Hawaii, Illinois, New Hampshire, and New York.

It is important to note that U.S. laws vary from state to state, but if someone is being threatened in e-mail, an online user may wish to ask a lawyer if federal statute 18 U.S.C. 875(c) might apply. Under that statute, transmission in interstate or foreign commerce of a communication containing any threat to kidnap any person or any threat to injure the person of another is a federal felony and carries a maximum prison sentence of five years and a fine of $250,000 (Hartman, 2006).

In April 1999, the first successful prosecution under California's cyber-stalking law occurred. Prosecutors received a guilty plea from a fifty-year-old former security guard who had used the Internet to encourage a sexual assault on a twenty-eight-year-old woman who had rejected the guard's romantic overtures. The charges were one count of stalking and three counts of soliciting sexual assault. The guard terrorized this woman by impersonating her in various Internet chat rooms and online bulletin boards, where he posted her telephone number, address, and messages saying that she fantasized about being sexually assaulted. On at least six occasions, sometimes in the middle of the night, men knocked on the target's door saying that they were there to meet her fantasies (Schell and Martin, 2004).

Apart from laws, a number of cyber-stalking resources exist online to assist targets in managing their situations and for getting protection, prevention, and recovery advice. These include CyberAngels (at http://www.cyberangels.org/), the world's oldest and largest Internet-safety organization, Coyote Communications (at http://www.coyotecom.com/), and SafeTeens (at http://www.safeteens.com/).

There are also some sound procedures that cyber-stalking targets can follow. First, they should report the incident to the system administrator of their Internet Service Provider (ISP) and that of the stalker or harasser. It is possible that the stalker will try to conceal his or her tracks by forging (that is, "spoofing") his or her e-mail header, but complaints of abuse are generally sent to the postmaster or the abuse department (with addresses something like postmaster@yourisp.com or abuse@yourisp.com). In fact, many ISPs use both addresses. After investigating the complaint, most system administrators will not hesitate to cancel the account of anyone using the ISP to send abusive e-mail (Hartman, 2006).

Anyone receiving abusive e-mail might want to visit Spam-Cop at http://www.spamcop.net/. This free service will analyze the unwanted e-mail to determine its point of origin, and it can generate a report that will be e-mailed to the appropriate system administrator. Internet users can also visit the CyberSnitch Website at http://www.cybersnitch.net/cybersnitch.htm to report Internet abuse or suspected cyber-terrorism—and have a report sent to the appropriate law enforcement agency.

Moreover, in recent years, and particularly because of the importance of obtaining credible cyber evidence for judges, police officers have had to get special training in cyber forensics to deal properly with Internet crimes. A reported cyber-stalking case not that long ago demonstrates how the lack of law enforcement training and expertise can cause psychological harm, if not physical harm, to a cyber-stalking target.

A mother complained to a local police agency that a man had been posting information on the Web, saying that her nine-year-old daughter was available for sex. The Web posting included their home phone number—with an invitation to telephone that number twenty-four hours a day. The family received numerous phone calls. Disgusted with this cyber-stalker, the family then went to the local police to report the problem and their rising fears. The police initially advised the family simply to change their home phone number. Bothered by this advice, the family then contacted the FBI, and they started an investigation. The FBI later discovered that the reason the family was given this poor advice was that the police did not have a computer expert with the appropriate training (U.S. Department of Justice, 2003).

Finally, anonymous services on the Internet—known as anonymous remailers—make cyber-stalking detection difficult for three reasons. First, anonymous services allow online users to create a free electronic mailbox through a Website. Second, while most entities providing this service request identifying information from online users, such services almost never actually authenticate the information. Third, for these services, a payment may be made in advance through the use of, say, a money order or other nontraceable form of payment. As long as the money is received in advance by the Internet Service Provider, the Web service can be provided to the unknown account holder.

In short, anonymous remailers purposefully strip identifying information and transport headers from electronic mail. By forwarding e-mails through several of these services serially, a cyber-stalker can transmit numerous quite anonymous harassing and threatening e-mails to targets. The availability of these services makes it relatively easy to send anonymous communications but relatively difficult for law enforcement to identify the person or persons responsible (ibid.).

A Case of Past and Present Online Gaming Controversies.   A series of events occurring in the late 1990s in a text-based online virtual community known as LambdaMOO produced much controversial discussion about whether "virtual rape" is or should be a criminal offense. These events also focused on whether cyber-stalking actually occurred in this virtual space.

In 1998, Julian Dibbell described the cyber "complaints" that grew in LambdaMOO (Dibbell, 1998, p. 1):

> They say he raped them at night. They say he did it with a cunning little doll, fashioned in their image and imbued with the power to make them do whatever he desired. They say that by manipulating the doll, he forced them to have sex with him, and with each other, and to do horrible, brutal things to their own bodies. I can assure you that what they say is true, because it all happened right in the living room—right there amid the well-stocked bookcases and the sofas and the fireplace—of a house I came later to think of as my second home.

Indeed, LambdaMOO was a Black Hat equivalent of the present-day popular White Hat online game, Sims Online. To be more precise, it was a subspecies of MUD (a multiuser dungeon) known as a MOO—short for "MUD, object-oriented."

It was a kind of database designed to give users the vivid impression of moving through a physical space. When users dialed into LambdaMOO, the program immediately presented them with a brief textual description of one of the database's fictional rooms in a fictional mansion. The rooms, the things in them, and the characters were allowed to interact according to rules roughly mimicking the laws of the physical world. In general, LambdaMOOers were given the freedom to create. They could describe their characters any way they liked, they could decorate the rooms any way they wanted, and they could build new objects. Although there was an illusion of presence, what the user really saw when visiting LambdaMOO was a kind of slow-moving script, lines of dialogue, and stage direction creeping steadily up the computer screen (ibid.).

On the night in question, the cyber-perpetrator was a LambdaMOO individual, known as Mr. Bungle, who used an online voodoo doll and a piece of programming code to spoof other players by appropriating their identities. In the context of LambdaMOO, this meant that by typing actions into the virtual voodoo doll, Bungle could make it appear as if another player in LambdaMOO was performing certain actions. One evening, Mr. Bungle logged into LambdaMOO and used the voodoo doll to make it appear that a number of the female participants were engaging in various forms of sexually humiliating activities. One player using the moniker Moonfire saw on her computer screen these obnoxious words: "As if against her will, Moonfire jabs a steak knife up her ass, causing immense joy. You hear Mr. Bungle laughing evilly in the distance." (Brenner, 2001, p. 27).

The targets of Mr. Bungle's attention were shocked and traumatized by how he had manipulated their characters and by how powerless they had been to stop him. Outraged by their suffering, some targets demanded "capital punishment" for Mr. Bungle, insisting that his character be annihilated. Others disagreed, claiming freedom of speech.

Before the issue was formally resolved, one member of the cyber community eliminated Mr. Bungle's persona and the corresponding user account in the system. Although Mr. Bungle's real-life puppeteer was a New York University computer user, it was clear that Mr. Bungle could not be prosecuted under the rape laws that existed at the time, as Mr. Bungle (and his real-life counterpart) did not commit the crime of rape—for that requires a physical assault. Moreover, while the LambdaMOO incident also had elements of identity theft—the malicious misuse of someone else's identity—participants in the cyber community did not allege any such infraction (Brenner, 2001).

Moving into this millennium, the New York University computer user might have been given a much harder time by the cyber community—and the legal system—for his cyber game exploits. While virtual-world owners have long faced difficulties in policing the security of their systems, combining sophisticated code exploits with virtual property markets can today be big business—sometimes resulting in charges being laid.

In South Korea in 2003, for example, a twenty-two-year-old student named Choi and an accomplice manipulated a virtual world server and got 1.5 billion won, or about 1.2 million U.S. dollars. According to reports, these two people were arrested amid claims that they had cracked an online game server and awarded themselves huge amounts of cyber money, which they later exchanged for real South Korean money. A spokesperson for South Korea's national police agency told the press that Internet criminals are becoming increasingly organized, and that they target online gaming sites in the hope of defrauding the services and converting cyber money into hard cash. At the time of this incident, 40,000 Internet crimes had been reported in South Korea—an 18 percent increase over the same period in 2002. More than 22,000 of these Internet crimes involved online gaming sites, which are extremely popular in South Korea (Wearden, 2003).

In the United States, the federal Computer Fraud and Abuse Act (CFAA) would seem to apply to such online gaming misdeeds. A criminal violation of Section 1030, in particular, requires three main elements—all of which are present in the Choi case:

- The defendant must have intentionally accessed a computer to commit a wrongdoing,
- The access must have been without authorization or exceeding the scope of the defendant's authorization,
- The damaging harm resulting from the unauthorized access must exceed $5,000.

Moreover, on July 30, 2003, the chief judge of a mock U.S. district court in Las Vegas, Philip M. Pro, heard *United States v.*

J.B. Weasel. Like Choi, Mr. Weasel was accused of violating the CFAA by attacking the servers of Getta Entertainment, maintainers of the virtual game GettaLife. Mr. Weasel allegedly directed a person who controlled an avatar called Terron to crack Mr. Martin's account in GettaLife, steal all of his virtual assets (especially his endeared "Staff of Viagra"), and leave his avatar unclothed and defenseless in the game. The case was entirely fictitious, of course. The moot was conducted at the 2003 Black Hat security conference of network and computer security experts (Lastowka and Hunter, 2006).

The interesting point in terms of online gaming's legal evolution, however, was that the contrived jury agreed with the prosecution's arguments that the virtual property at issue had real value—and that real harm could, therefore, be caused to persons or property. Edward Castronova, an expert witness for the prosecution, reported (ibid., p. 314):

> Defense counsel Jennifer Granick mounted a strong counterargument, namely that we might, as a society, decide that it is just too difficult to classify game-related damages as real, just as we shy away from taking cases of lost sexual favors to court, even though there clearly are damages. This powerful argument suggests that losses in something we agree to call a "game" should also be free from legal oversight, even though, in fact, the distinction between game and life is arbitrary. In the end, jury and audience disagreed with this cultural stratagem, preferring instead Prosecutor Richard Salgado's argument that human activity in the allegedly virtual space is not virtual at all. It is real activity and has real values and thus, in principle, it deserves the full attention of policy and law.

If such a finding actually occurs in a real courtroom, it would be an obvious first step in North America toward the legal recognition of virtual property, and online users who could possibly benefit from such an argument would be online game owners victimized by commercial exploiters.

An example would be a gold dupe, where an online player would, by exploiting game code, generate duplicate currency. If the player had sufficient game account and computers exploiting this type of dupe, it would be possible to create so much excess gold that the virtual currency would be devalued in the game. For the exploiter, a dupe can generate a large number of real-world dollars before the exchange rate drops drastically. This sort of crime is like counterfeiting, because dupers and exploiters are trying to create value illegitimately, but the creation of such surplus currencies can destroy functioning economies by creating frustration in online gamers, causing them to stop their online game subscriptions. In short, game owners who are victims of such commercial exploiters could point to real economic harm created by the unauthorized access and sale of virtual currencies. These harms may end in criminal prosecution for property crimes through the Internet, running counter to computer trespass statutes (Lastowka and Hunter, 2006).

Using similar arguments, if the real-life person were to conduct the virtual abuses of Mr. Bungle in LambdaMOO in 2006, the real-life university-computer user might face a conviction. The argument could be made that he caused psychological harm to persons in this virtual community, totaling in excess of $5,000.

Present-day Identity Theft Complaints. Although no one in the virtual community complained about Mr. Bungle as an identity theft case, today, identity theft, both in the real world and on the Internet, is one of the most troubling and increasingly harmful cyber-crimes.

The malicious theft of and consequent misuse of someone else's identity to commit a crime is known as identity theft or masquerading. Identify theft often involves cracking into a system to obtain personal information—such as credit card numbers, birth dates, and social security numbers—and then using that information in an illegal manner. The cracker may buy items with the stolen identity or pretend to be someone else of higher professional status to gain special privileges. Identity theft is one of the fastest growing crimes in the United States and elsewhere around the globe.

Often identity theft victims are informed by their financial institutions or places of employment that there has been an exploit in the computer network, leaving their personal information at risk. At other times, victims often suddenly discover that someone has stolen their identities, emptied their bank accounts, spent on their credit card accounts to the maximum limitation, and left them with a huge debt to pay to clear their credit ratings.

Sometimes, tragically, the impostor has committed a serious crime using the victim's identity, leaving the victim with an undeserved criminal record. Although identity theft is often viewed as a high-tech, Internet-assisted crime, the thief can be, and often is, a family member, a trusted friend, or a coworker having knowledge of the target's personal information, including account passwords (Hammond, 2003).

Since 1999, an alarming number of Internet Website cracks, personal information leaks, and potential identity theft incidents have been reported in the news media. For example, in February 2005, a report released by a U.S. legislature committee found that information on the Websites of the New York Department of Motor Vehicles, the Department of Education, the Department of Correctional Services, the State Division of Military and Naval Affairs, and the New York Power Authority had been defaced a whopping seventy-two times from 1999 to early December 2004. The chair of the oversight committee investigating the problem said that because state and private companies are not keeping important personal and homeland security online information safe, identity thefts could occur.

Moreover, in March 2005, more evidence emerged of the poorly regulated Internet information brokerage industry. About March 10, cyber thieves stole passwords from legitimate online users on as many as 32,000 Americans in a database owned by the trustworthy LexisNexis Group (affecting 310,000 clients). Similar computer network breaches occurred at ChoicePoint, Inc., and the Bank of America (affecting about 1.2 million federal employees with this charge card).

These intrusions prompted calls for the U.S. government to investigate this serious set of events. Government officials agreed to review the online security policies of the Bank of America, as well as Bank One of Delaware, Citibank of New York, Mellon Bank of Pittsburgh, and the U.S. Bank of Minneapolis. Recommendations were to be forthcoming about how to provide adequate protection for the safety of federal employees' personal information.

Although the first in a series of hearings was scheduled to begin on Capitol Hill on March 10, 2005, the U.S. states are collectively proposing more than 150 bills to regulate online security standards, to increase identity theft and fraud protection, to increase data broker limitations, to increase limits on data sharing or use or sales, and to improve security breach notification.

In recent years, some seemingly innocent-looking techniques have been used by malicious online users to commit identity theft crimes. Phishing is a form of identity theft, whereby a scammer uses an authentic-looking e-mail from a large corporation to trick e-mail receivers into online disclosure of sensitive personal information, such as credit card numbers or bank account codes.

According to a 2004 report released by Garner, Inc., an IT marketing research firm, in 2003, phishing exploits cost banks and credit card companies an estimated $1.2 billion. Moreover, according to the Anti-Phishing Working Group (a nonprofit group of government agencies and corporations trying to reduce cyber fraud), more than 2,800 active phishing sites are known to exist on the World Wide Web.

In April 2005, a new "cousin" of phishing was defined and called "WiPhishing" (pronounced "why phishing")—an act executed when an individual covertly sets up a wireless-enabled laptop computer or access point to get other wireless-enabled laptop computers to associate with it before launching a crack attack. Because about 20 percent of wireless access points use default SSIDs (like Linksys), a cracker can quite easily guess the name of a network that target computers are normally configured to, thereby gaining access to the laptop computer. Once access is gained, crackers can put malicious code into it. Intrusion detection appliances like AirPatrol Enterprise have been designed to detect wireless exploits.

Firms having wired networks are at risk of being cracked if employees' laptop computers are left on. Instead of having crackers exploit wireless networks with WiPhishing, they could do even more damage by hijacking the legitimate connection to a wired computer network, exploit the soft underbelly of that network, and launch an invasive attack (Leyden, 2005).

Similar in nature to phishing, pharming is done by crackers wanting to get personal or private (usually financially related) information through domain spoofing. So, instead of spamming targets with ill-intended e-mail encouraging them to visit spoof Websites appearing to be legitimate, pharming actually poisons a DNS server by putting false information into it. The result is that the online user's request is redirected elsewhere. Often the online user is unaware that this is happening, because the browser indicates that the user is at the correct Website. Because of this outcome, Internet security experts view pharming to be a more serious exploit, because it is more difficult to detect. In short, while phishing exploits try to scam targets on a one-on-one basis, pharming lets Black Hats scam large numbers of targets all at once by effectively using domain spoofing.

It is interesting to note that on March 4, 2005, White Hat hackers surfed the Web at Seattle University with the intention of harvesting Social Security and credit card numbers—a replication of what goes on in the real world by Black Hats. In less than sixty minutes they found millions of names, birth dates, and Social Security and credit card numbers using just one Internet search engine, Google. They warned that the use of the right kind of sophisticated search terms could even find data deleted from company or government Websites but temporarily cached in Google's extraordinarily large data warehouse. The problem did not lie with Google, they affirmed, but with companies allowing Google to enter into the public segment of their networks (called the DMZ) and to index all the data contained there. Although Google does not need to be repaired, said the White Hats, companies and government agencies need to understand that they are exposing themselves and their clients by posting sensitive data in public places (Shukovsky, 2005).

Present-day Fraud Complaints.   Because of its often anonymous and decentralized composition, the Internet is fertile ground for fraud exploits. Fraud is generally defined in law as an intentional misrepresentation of facts made by one person to another person, knowing that such misrepresentation is false but will, in the end, induce the other person to act in some manipulated fashion that results in injury or damage to the person manipulated.

Fraud may include an omission of facts or an intended failure to state all of the facts. Knowledge of the latter would have been needed to prevent the other statements from being misleading. In cyber terms, spam is often sent in an effort to defraud another person into purchasing something that he or she has no intention of purchasing. Fraud can also occur through online gaming, as noted earlier, or through online auctions. Fraud can also occur in online voting.

Recently in the United States, the Sarbanes-Oxley Act (SOA) was passed as a reaction to the accounting misdeeds in companies such as WorldCom and Enron, but its passage has fraud implications for online personal information storage. Simply stated, with the vast amounts of personal information stored on company computers, fraud opportunities abound for cyber-criminals. A major problem prompting the passage of this act was that companies storing huge amounts of information have tended to give little thought to what is being stored in company or institutional networks, or how securely it is being stored. Consequently, occasional occurrences of fraud or alterations of data by crackers have often gone undetected.

Experts have argued that, rather than spending lots of money to store data in accordance with the act, companies should allocate some money to determine exactly what kinds of information need to be stored and for how long. Many companies have policies, for example, dictating that data be stored for periods lasting from six to nine months, but that timeline may not be realistic. Such confusion over this important information storage issue may be a primary reason why the Sarbanes-Oxley Act deadline for companies based in European countries has been pushed back another year. Originally, controversial Section 404 of the SOA outlined the requirement for companies to archive information by July 15, 2005.

And on an optimistic note, in the fight against online auction fraud, consumers fed up with being ripped off by online criminals have started to fight back through cyber-vigilantism. They use the Internet to get even with cyber thieves, and they band together to inform federal authorities of their online plights.

This interesting story started on or about August 10, 2002, when a man named Mark was the highest online auction bidder for a Toshiba Protege 2000 laptop computer, listed on eBay. He sent a cashier's check for $1,485 to the eBay seller Tech-Surplus, but by September 1, 2002, he had still not received the item. Frustrated, Mark e-mailed other winning bidders to see if their items had been received from Tech-Surplus (Steiner, 2002). His worries were confirmed. No one had received the purchased computers. One of the e-mail respondents, named Karen, volunteered to create a Website to publish revealing information about the seller, who was suspected of acting in bad faith. Although originally set up on GeoCities, the vigilante Website was later moved to http://www.techsurplusvictim.com (ibid.). As of February 2, 2006, this site was still active.

Another respondent named Cory created a mailing list on Yahoo Groups to arrange for an online communications network for Tech-Surplus targets. With this effort, Mark, Karen, and Cory soon realized that they were just three of a group of more than eighty online targets who were similarly hit by Tech-Surplus. Although Karen and Cory received refund checks from the seller in what they perceive was an attempt to get them to cease and desist from their cyber-vigilante activities, the attempt failed. Karen continues to put in about six hours a day to keep the Website active. She also contacts law enforcement officials and speaks about the Website's findings to the press (ibid.).

Finally, to deal with cyber incidents resulting in fraud and identity theft, the Internet Fraud Complaint Center (IFCC) was created in the United States in May 2000 by the FBI and the National White Collar Crime Center. The IFCC deals with complaints about Internet fraud and provides targets with resources for protecting themselves. To date, the IFCC has received thousands of complaints weekly about crimes including online auction frauds, undelivered goods, credit card fraud, identity theft, and nonpayment for services.

This agency is now called the Internet Crime Complaint Center, or IC3, and is found at Website http://www.ic3.gov/. IC3's mission is to serve as a vehicle to receive, develop, and refer criminal complaints to authorities involving the Internet. Since its creation, the IC3 has received complaints crossing a wide range of Internet crimes, including online fraud in many forms, Intellectual Property Rights matters, computer intrusions (that is, cracking), economic espionage (that is, stealing trade secrets), online extortion, international money laundering, and identity theft. Since June 2000, the IC3 has realized that regardless of the label placed on a particular Internet crime, the potential for it to overlap with another referred matter is substantial.

Since 2002, a number of U.S. states (as well as other jurisdictions) with high online fraud complaint records have been fighting back through improved legislation. For example, California Senate Bill 1386 was signed into law on September 25, 2002. It amended the California Civil Code to require that notice be given to the Department of Defense (DoD) regarding security breaches involving unencrypted (that is, unprotected) personal information.

Cyber-pornography and Child Cyber-pornography Complaints. Because of the relative anonymity offered by the Internet and the users' capacity to create and amend visuals using online graphics and software technology, the prevalence of cyber-pornography—and child cyber-pornography, in particular—has soared over the past decade. Without question, they present major moral and personal harm problems for society.

For example, TopTenReviews, Inc., estimates that online child pornography, alone, generates about U.S.$3 billion annually. More than 100,000 Internet Websites currently offer child pornography. What is frightening is that the demand for online child pornography appears to exceed that for adult pornography, relative to supply. According to this same review, online adult pornography generates about U.S.$2.5 billion annually. In total, more than 4 million pornographic Websites—about 12 percent of all Websites—exist (Ropelato, 2006).

In the United States, child pornography is a category of speech not protected by the First Amendment. The federal legal definition of child pornography can be found at 18 U.S.C. § 2256. Some particulars around the definition have changed in recent years, with the latest change occurring on April 30, 2003, when President George W. Bush signed the PROTECT Act.

The latter not only implemented the Amber Alert communication system—which allows for nationwide alerts when children go missing or are kidnapped—but also redefined child pornography to include both images of real children engaging in sexually explicit conduct and computer-generated depictions indistinguishable from real children engaging in such acts. *Indistinguishable* was further defined as that which an ordinary person viewing the image would conclude is a real child engaging in sexually explicit acts. However, cartoons, drawings, paintings, and sculptures depicting minors or adults engaging in sexually explicit acts, as well as depictions of actual adults that look like minors engaging in sexually explicit acts, are excluded from the definition of child pornography.

The PROTECT Act of 2000 was passed because of three major problems that remained despite previous legislation (U.S. Department of Justice, 2003). These were as follows:

- Law enforcement did not have the tools needed to locate missing children and to prosecute offenders.
- Existing federal laws did not ensure adequate, and at times, consistent punishment for those found guilty of such crimes.

- Past legal obstacles have made prosecuting child pornography cases very difficult—especially in the case of virtually produced child pornography.

Prior to the enactment of the PROTECT Act, the definition of child pornography came from the 1996 Child Pornography Prevention Act (CPPA). Moreover, the Children's Online Privacy Protection Act (COPPA), effective April 21, 2000, applied to the online collection of personal information from children under the age of thirteen. The rules detailed what a Website operator must include in a privacy policy, when and how to seek verifiable consent from parents or guardians, and what actions an operator must take to protect children's privacy and safety online. It is important to note that these Internet safety policies required the use of filters to protect against access to visual depictions considered obscene or harmful to minors (Miltner, 2005).

A filter is a device or material that suppresses or minimizes waves or oscillations of certain frequencies. Therefore, filtering software should block access to Internet sites listed in an internal database of the product, block access to Internet sites listed in a database maintained external to the product itself, block access to Internet sites carrying certain ratings assigned to those sites by a third party or that are unrated under such a system, and block access based on the presence of certain words or phrases on those Websites. In short, software filters use an algorithm to test for the appropriateness of Internet material—in this case, for minors.

Websites are first filtered based on IP addresses or domain names. Since this process is based on predefined lists of appropriate and inappropriate sites, experts suggest that relying totally on these lists is ineffective because Internet sites come and go so quickly. Moreover, though minors often frequent online chat rooms, instant messaging, and newsgroups, those are not under the filtering system but are a real point of concern for society (ibid.).

It is important to note that librarians in the United States have had particular difficulty enforcing the use of filters and the requirements of CIPA, because they are faced daily with both protecting minors from viewing obscene materials and also not obstructing adults from viewing materials they are legally allowed to obtain and peruse under the First Amendment, offensive or not. To that end, in the 2002 case *American Library Associa-*

tion v. the United States, the District Court for the Eastern District of Pennsylvania held CIPA to be unconstitutionally broad because of its crude approach to filtering technology. Furthermore, the district court noted that any technology protection measure mandated by CIPA will necessarily block access to a large amount of speech whose suppression serves no legitimate government interest. Besides determining that CIPA's content-based filtering restrictions were subject to tight scrutiny, the district court also considered whether the libraries' use of filters to protect minors constituted a legitimate state interest. In the end, the district court saw the Internet as a public forum—much like sidewalks and parks—and ruled that the First Amendment should, therefore, be promoted in this public forum. Also, the district court concluded that less restrictive use policies should be implemented by librarians rather than adhering to CIPA's overly inclusive filtering systems (ibid.).

Nevertheless, on June 20, 2002, the U.S. government appealed the district court's ruling, and on June 23, 2002, the U.S. Supreme Court reversed the district court's decision. The Supreme Court upheld CIPA's constitutionality on the grounds that this piece of legislation does not violate the First Amendment, because U.S. public libraries do not offer Internet access to produce a public forum for Web publishers but to facilitate research, learning, and the recreational pursuits of library patrons by providing materials of appropriate quality. The Supreme Court held that librarians need to select books and materials for public shelves on a rational basis rather than using strict scrutiny. With regard to the filters and their tendency to overblock First Amendment–protected speech, Chief Justice Rehnquist affirmed that CIPA allows librarians to disable a filter without significantly delaying adult users' requests to view material, and that protecting library minors from viewing material inappropriate for them outweighs any temporary inconvenience to adult library patrons (*United States v. American Library Association,* 2003).

Following that Supreme Court ruling, U.S. librarians argued that though CIPA allows but does not require librarians to disable Internet-filtering software, there are no procedures in place to guide librarians as to the proper circumstances under which disabling a filter is appropriate. Therefore, the disabling provision of CIPA will be applied unequally to patrons in various U.S. jurisdictions.

Certainly, regardless of existing laws, controlling what obscene materials minors view is a problem worldwide. For example, in 2002, Kinderconsument, a Dutch consumers' organization for children, published their own sampling study among 1,300 children from ten to thirteen years of age. The results were alarming. Up to 44 percent of the children—depending on the chat rooms they frequented—were bothered by floods of abuse and sexual intimidation. The children were also disgusted to find that annoying men who turned out not to be children at all flooded them with abusive comments and words. They received unwanted pornographic pictures online, along with invitations to meet in person. Moreover, there seemed to be gender differences in terms of coping with the stress in chat rooms. Girls admitted to experiencing emotionally complex and disturbing situations on the Internet, but they tended not to tell their parents. Some even admitted to having an online love affair—without their parents knowing (Benschop, 2005).

Royal Mounted Canadian Police corporal Jim Gillis, head of Project Horizon, a policing initiative dealing with online child pornography and based in Halifax, Nova Scotia, said that home PC owners and businesses play an unknown but key role in promoting such criminal activities, for a large part of the problem arises from the fact that "bots" are often planted by a virus on home and business computers to convert them into zombies that are remotely controlled by cyber-criminals. Although the computers appear to be operating normally, they could actually be relaying child pornography traffic or storing child porn images. In this way, the cyber-criminals can avoid detection (Butters, 2005). To avoid being part of the criminal chain, PC users and businesses should have antivirus software on their computers, as well as firewall and network protection.

In a report on children as victims of violent crime released on April 20, 2005, the Office of Statistics Canada said that charges related to child pornography increased eightfold over the period from 1998 through 2003. The increase in charges laid by law enforcement agents in Canada has been a result of several factors, including increased public awareness about the potential of the Internet to cause harm to children, police having increased resources to conduct the investigations, and improvements in technology. Also, volunteers patrolling the Internet for cyber-criminals have aided law enforcement in their capture.

In 1995, the CyberAngels started to appear online to patrol the World Wide Web around the clock in the battle against child pornography and cyber-stalking. In 1999, the organization helped to locate child pornography sites, resulting in the first-ever set of arrests in Japan of Internet child pornographers (Karp, 2000).

The frightening reality is that at least 80 percent of those who purchase child pornography are active child molesters. Moreover, 36 percent of child pornographers who use the U.S. mail to exploit a child have been found to be actual child molesters. Child pornographers range in age from 10 to 65 (Posey, 2005).

Suspected child pornography Websites can be reported online at www.Cybertip.ca. The latter is operated by Child Find Manitoba and launched in Canada at the end of January 2005. Also, as of February 2005, a Child Exploitation Tracking System went into operation in Canada, made available by Microsoft founder Bill Gates. The Child Exploitation Tracking software helps police share information about cyber-pornographers by streamlining the difficult task of managing huge information stores.

In recent years, a number of technological attempts to combat cyber-pornography have been developed. First, there are host-based solutions such as CBIR—content-based image and video retrieval. This area is very important because of emerging child-pornography employing both image and video files. The objective is to retrieve visual files based on their semantics. The semantics of a file are determined according to a set of characteristics like color contrast and shapes "learned" from similar files (Chopra et al., 2006).

Second, there are censured P2P solutions. A number of P2P program vendors have recently entered alliances to combat child pornography. Although most programs used for file sharing have not in any way been part of this alliance (such as the Japanese software named Winny—a Japanese P2P file-sharing program claiming to be inspired by the Freenet network, and that also claims to be able to keep online users' identities untraceable), the state of P2P programs is quickly changing to include heavy encryption and anonymity features (ibid.; Wikipedia, 2006).

Third, child pornography can be combated at the network level. The technical approaches can be classified within a traffic-type model consisting of visual-type traffic, text-type traffic, and encrypted-type traffic. Visual-type traffic consists of moving pictures or frames. Text-type traffic consists of things like e-mail and documents. Encrypted-type traffic represents a more challenging phenomenon, since it may be computationally unfeasible to analyze heavily encrypted file contents.

Fourth, a number of new classification systems are under development. One of these is a classifier combining a stochastic learning-based weak estimator and a linear classifier, so that data in each IP packet of 512 bytes drawn from a source alphabet can be classified as "obscene" or "nonobscene." If this classification system meets the industry standard, it should be able to classify both textual and visual child pornography (Chopra et al., 2006).

Violations of Intellectual Property Rights.

Infringing Intellectual Property Rights (IPR) and copyright can occur online and, thus, fall in the broad-based category of "cyberspace theft." An example of such infringement would be copying another's work from an online source without being authorized to do so or without giving credit for the work—including songs, articles, movies, or software.

In January 2000, one of the interesting cases to make headlines in the United States was the Internet free speech and copyright civil court case involving *2600: The Hacker Quarterly*, Universal Studios, and members of the Motion Picture Association of America. Here, legal issues emerged around *2600's* alleged violation of the Digital Millennium Copyright Act (DMCA). Enacted in October 1998, the DMCA was intended to implement under U.S. law certain worldwide copyright laws to cope with emerging digital technologies by providing protection against the disabling or the bypassing of technical measures designed to protect copyright. DMCA sanctions apply to anyone who attempts to impair or disable an encryption device protecting a copyrighted work, typically using the Internet. A copy of the DMCA can be found at this Website: http://www.copyright .gov/legislation/dmca.pdf.

The problem arose for *2600* when, in November 1999, the hacker vehicle linked to and discussed a computer program called DeCSS, DVD decryption software. The complainants objected to the publication of DeCSS, because they argued that it could be used as part of a process to infringe copyright on DVD movies. However, in their defense, *2600* claimed that decryption of DVD movies is necessary for a number of reasons, including to make "fair use" of movies. In the end, the hacker magazine lost the case, despite the excellent legal advice and representation by the EFF.

Another interesting court case surrounding Intellectual Property Rights (IPR) arose in 2002, but the events around that case were really quite bizarre. Seldom does a new product land in court before it is displayed on store shelves. But that was the case with DVD X Copy, a newly developed software program that gave consumers who purchased a DVD a relatively easy way to burn a backup copy of a prerecorded, copy-protected DVD movie for safekeeping but without breaking copyright laws. What is even more unusual in this particular incident is that the lawsuit was started by the software program's creator, 321 Studios, Inc. The reason for the company-driven court case was that the company wished to obtain a definitive ruling from the U.S. courts on whether making one's own personal backup copies of DVDs—without sharing them with friends—is a legal activity under U.S. copyright laws. Despite its neophyte character, DVD X Copy raised considerable controversy regarding the murky line between the protection of copyright laws and consumer fair use rights, particularly when most types of entertainment are becoming digitalized (Evangelista, 2002).

Another controversial issue regarding IPR came to light with the release of Sony's Digital Rights Management (DRM) software used on some of its CDs—a content protection scheme. Purchasers who, say, bought the Sony BMG's "Get Right with the Man" CD by the Van Zant brothers may have noticed that it has been protected with DRM. What this means in real terms is that the CD can be played only using the media player that ships on the CD itself, and that purchasers are limited to at most three copies. Some consumers became angered not only that Sony had put software on their systems that utilizes techniques commonly employed by malware to mask its presence but also that the software is so poorly written that it provides consumers no way of uninstalling it. To make matters worse, most consumers who have purchased the DRM CD and who discover the cloaked files with, say, an RKR (RootkitRevealer) scan, will cripple their computers if they try to delete the cloaked files. Although most consumers would probably uphold the media industry's right to employ copy-protection devices to curb illegal CD copying, many would argue that, to date, society has not found the right balance between fair use and copy protection—and that Sony's employment of DRM may be taking IPR protection one step too far (Mark's Sysinternals Blog, 2005).

Without question, the social issue of infringing Intellectual Property Rights, copyrights, and patents has drawn considerable debate from those who fight for freedom of information as well as from those who fight against abuses of artists' rights and royalties. For this reason, during the 2004 U.S. presidential campaign, the INDUCE Act, or Inducing Infringement of Copyright Act of 2004, was proposed by Senator Hatch of Utah.

If passed, the latter could have killed the market for digital music devices like Apple iPods—which copy music from users' computers. The INDUCE Act would have criminalized digital music technologies because they could be viewed as inducing others to infringe copyright. When news about the INDUCE Act surfaced, hacktivists (that is, online political activists) went to work, constructing Websites like www.Savetheipod.com to motivate music lovers to send letters of opposition to Congress. The electronics industry and the Electronic Frontier Foundation (EFF) also lobbied against its passage. The INDUCE Act met its demise in October 2004, but, if passed, this far-reaching piece of legislation could have forced electronic companies and Internet services to get permission for each new technology developed.

A more recent legal case regarding IPR and Internet technology made the headlines in 2006. This fascinating case—which ended with a negotiated settlement of U.S.$612.5 million in March 2006—involved two high-tech companies—RIM, Ltd. (Research in Motion, a Canadian company based in Waterloo, Ontario) and NTP, Inc. (a small U.S. company based in Virginia). Each company claimed that it owned the Intellectual Property and patents for the wireless handheld used by business and government people around the world to read and respond to their e-mail while away from their computers.

Named "BlackBerry" by RIM, these nifty little handhelds had been called "Crackberries" because users tended to become addicted to using them. The devices became so popular that, during the legal proceedings, the U.S. federal government requested that employees with mission-critical roles be exempted from an injunction, should one be levied by the courts (Arnone, 2006).

On February 1, 2006, RIM, Ltd., secured a key victory after U.S. patent authorities rejected the last of five contested patents at the heart of the dispute, and immediately RIM shares rose nearly 9 percent on the Toronto Stock Exchange, as investors bet that RIM was gaining the edge in the five-year patent saga. However, with the negotiated settlement, the CEO of RIM agreed to pay NTP a one-time payment of over U.S.$600 million to receive a license to NTP's patents going forward.

This case, and others of a similar nature, clearly show how cumbersome and costly patent disputes can become. Estimates at one point were that a possible settlement could place at over U.S. $1 billion. (McKenna, 2006).

Conclusion

Indeed, while the Internet has created many opportunities for society to improve the lifestyle of its citizens, costs have also arisen. As in the noncyber world, for every positive outcome that flows from real-world events and discoveries, there are counterbalancing negative outcomes committed by criminals with narcissistic motives. The same can be said for the virtual world.

Chapter 2 has focused on the controversies surrounding the Internet, particularly those of a privacy, security, and trust nature. A number of crimes perpetrated by Internet abusers were detailed, as were the legal, social, and technological solutions. With the evolution of the Internet and with its increased globalization, new social issues will continue to emerge, and novel legal, social, and technological solutions will need to be found for resolving them. The growth of the Internet in developing countries will be the focus of Chapter 3.

References

Al-Fedaghi, Sabah. "How to Calculate the Information Privacy." University of New Brunswick, http://www.lib.unb.ca/Texts/PST/2005/pdf/fedaghi.pdf (accessed October 12–14, 2005).

Arnone, Michael. "Contractors Told to Relax about Blackberry." 101 Communications, http://www.fcw.com/article92023–01–23–06-Print (accessed February 2, 2006).

Associated Press. "Business Schools: Harvard to Bar 119 Applicants Who Hacked Admissions Site." *Globe and Mail,* March 9, 2005, p. B12.

Benschop, Albert. "Child Pornography in Cyberspace." SocioSite, http://www2.fmg.uva.nl/sociosite/websoc/pornography_child.html (accessed December 28, 2005).

Brenner, Susan. "Is There Such a Thing as 'Virtual' Crime?" Susan Brenner's Website, http://www.crime-research.org/library/Susan.htm (accessed 2001 in the California Criminal Law Review).

Butters, George. "Criminal Activity: Your Computer May be Housing Child Porn." *Globe and Mail,* January 27, 2005, p. B14.

Center for Education and Research in Informance Assurance and Security (CERIAS). "Info about SATAN." Purdue University, http://www.cerias.purdue.edu/about/history/coast/satan.php (accessed June 2, 1995).

Chopra, M., et al. "A Source Address Reputation System to Combating Child Pornography at the Network Level." IADIS International Conference on Applied Computing, February 25–28, 2006, San Sebastian, Spain.

Cline, Jay. "The ROI of Privacy Seals." ComputerWorld, Inc., http://www.computerworld.com/developmenttopics/websitemgmt/story/0,10801,81633,00.html (accessed June 2, 2003).

Dibbell, Julian. "A Rape in Cyberspace (Or TINYSOCIETY, and How to Make One)." Julian Dibbell Website, http://www.juliandibbell.com/texts/bungle_print.html (accessed 1998).

Evangelista, Benny. "Burning Debate: New DVD-copying Software Sets Stage for Legal Showdown with Hollywood." *San Francisco Chronicle,* http://www.sfgate.com/cgi-bin/article.cgi?f=/c/a/2002/12/09/BU141167.DTL (accessed December 9, 2002).

Flinn, S., and J. Lumsden. "User Perceptions of Privacy and Security on the Web." University of New Brunswick, http://www.lib.unb.ca/Texts/PST/2005/pdf/flinn.pdf (accessed October 12–14, 2005).

Grami, A., and B. Schell. "Future Trends in Mobile Commerce: Service Offerings, Technological Advances, and Security Challenges." University of New Brunswick, http://dev.hil.unb.ca/Texts/PST/pdf/grami.pdf (accessed October 13–15, 2004).

Hammond, Robert. *Identity Theft: How to Protect Your Most Valuable Asset.* New Jersey: Career, 2003.

Hartman, Rachel. "Cyberstalking and Internet Safety FAQ." Science Fiction and Fantasy Writers of America, Inc., http://www.sfwa.org/gateway/stalking.htm (accessed February 1, 2006).

Jonczy, J., and R. Haenni. "Credential Networks: A General Model for Distributed Trust and Authenticity Management." University of New

Brunswick, http://www.lib.unb.ca/Texts/PST/2005/pdf/jonczy.pdf (accessed October 12–14, 2005).

Karp, Hal. "Angels On-Line." *Reader's Digest* 157 (2000): 50–56.

Lastowka, F., and D. Hunter. "Virtual Crimes." New York Law School, http://www.nyls.edu/pdfs/lastowka.pdf (accessed February 1, 2006).

Leyden, John. "WiPhishing Hack Risk Warning." *Register*, http://www.theregister.co.uk/2005/04/20/wiphishing (accessed April 20, 2005).

Libbenga, Jan. "Classified Military Documents Found on P2P Site." *Register*, http://www.theregister.co.uk/2005/01/30/dutch_classified_info_found_on_kazaa/ (accessed January 30, 2005).

Mark's Sysinternals Blog. "Sony, Rootkits and Digital Rights Management Gone Too Far." Sysinternals, http://www.sysinternals.com/blog/2005/10/sony-rootkits-and-digital-rights.html (accessed October 31, 2005).

McKenna, Barrie. "RIM Wins Key Battle in Patent Dispute." *Globe and Mail,* February 2, 2006, pp. B1, B8.

Miltner, Katherine. "Discriminatory Filtering: CIPA's Effect on Our Nation's Youth and Why the Supreme Court Erred in Upholding the Constitutionality of the Children's Internet Protection Act." FindArticles, http://www.findarticles.com/p/articles/mi_hb3073/is_200505/ai_n15014919 (accessed May, 2005).

Parker, Donn. "Is Computer Crime Real?" California Criminal Law Review, http://www.boalt.org/CCLR/v4/v4letterstoeditor.htm (accessed 2001).

Posey, Julie. "Child Pornography: Is It So Bad?" Owl Publishing, http://www.pedowatch.com/porn.htm (accessed December 29, 2005).

Prescription Pricing Authority. "What Are Security Zones?" Prescription Pricing Division, http://www.ppa.org.uk/help/www/int00290.htm (accessed February 1, 2006).

Queeg Company. "Security Policy Checklist." Queeg, http://queeg.com/~brion/security/secpolicy.html (accessed October 6, 1997).

Ropelato, Jerry. 2006. "Internet Pornography Statistics." Top Ten Reviews, Inc., http://internet-filter-review.toptenreviews.com/internet-pornography-statistics.html (accessed 2005).

Schell, B., and N. Lanteigne. *Stalking, Harassment, and Murder in the Workplace: Guidelines for Prevention and Protection.* Connecticut: Quorum, 2000.

Schell, B., and C. Martin. *Cybercrime: A Reference Handbook.* California: ABC-CLIO, 2004.

Seddigh, N., et al. "Current Trends and Advances in Information Assurance Metrics." University of New Brunswick, http://dev.hil.unb.ca/Texts/PST/pdf/seddigh.pdf (accessed October 13–15, 2004).

Seigneur, J-M., et al. "Combating Spam with TEA (Trustworthy Email Addresses)." University of New Brunswick, http://dev.hil.unb.ca/Texts/PST/pdf/seigneur.pdf (accessed October 13–15, 2004).

Shukovsky, Paul. "Good Guys Show Just How Easy It Is to Steal ID." *Seattle-Post Intelligencer,* http://seattlepi.newsource.com/local/214663_googlehack05.html (accessed March 5, 2005).

Steiner, Ina. "eBay Auction Fraud Spawns Vigilantism Trend." Steiner Associates, http://www.auctionbytes.com/cab/abn/y02/m10/i12/s01 (accessed October 12, 2002).

United States v. American Library Association, 539 U.S. 194 (2003).

U.S. Department of Justice. "Fact Sheet. PROTECT Act." U.S. Department of Justice, http://www.usdoj.gov/opa/pr/2003/April/03_ag_266.htm (accessed April 30, 2003).

_____. "1999 Report on Cyberstalking: A New Challenge for Law Enforcement and Industry." U.S. Department of Justice, http://www.usdoj.gov/criminal/cybercrime/cyberstalking.htm (accessed February 7, 2003).

Wearden, Graeme. "Cybercriminals Target Online Gamers." CNET Networks, Inc., http://news.zdnet.co.uk/internet/security/0,39020375,39115585,00.htm (accessed August 8, 2003).

Wikipedia. "Winny." Wikimedia Foundation, Inc., http://en.wikipedia.org/wiki/Winny (accessed March 21, 2006).

3

Worldwide Perspective

The Internet has permitted Internet users around the globe to communicate with one another with the click of a button, entertain themselves with online gaming, enter votes for political candidates without leaving their home, and purchase products online—with store-to-door delivery. In fact, in 2001 the Forrester Market Research group predicted that worldwide electronic commerce (that is, e-commerce) would be a success to the tune of $6.8 trillion by 2004, and that growth in e-commerce was expected to continue at about a 5 percent rate over the next five years, particularly in countries in which citizens had readily available Internet access (McCarthy, 2001).

As consumer acceptance and advertiser ambitions have driven the rapid growth of the Internet, there has been considerable optimism regarding consumer interest in Web-delivered advertising and marketing tactics (Nowak et al., 1999). Studies of online users have reported that at least one-third of interactive households use the Web to investigate or buy products or services, with as many as 70 percent of regular Web users having made one or more online purchases in the recent past (Magill, 1998).

Besides allowing for growth in electronic commerce, the Internet has made possible enormous growth in the online educational market, both in developed countries and in developing countries. On April 4, 2001, the Massachusetts Institute of Technology (MIT) made headlines when officials announced that over the next decade, materials for nearly all of the courses offered at the university would be freely available on the Internet. This move was apparently inspired by the so-called White Hat spirit that has been the primary driving force behind the free information sharing movement found in the Computer Underground since the 1960s.

In North America, children from the early years on through graduate study tend to be connected to the Internet, making lifelong learning a reality for those even in rural areas. A survey in March of 2006 found that more than 46,000 schools for students between kindergarten and grade 12, community colleges, libraries, and museums in at least thirty-five U.S. states are now connected to the Internet2 backbone network. The survey also indicated the success of the nationwide Internet2 K20 Initiative in bringing together Internet2 member institutions and creative educators from elementary and secondary schools, colleges, universities, libraries, and museums to extend new technologies, applications, and advanced content to all educational sectors across the United States (Rotman, 2006).

The survey results also showed that 37 percent of all U.S. K–12 schools, 54 percent of all U.S. community colleges, 57 percent of all U.S. four-year colleges, and 20 percent of all U.S. libraries are now connected to Internet2's nationwide network. Equally as impressive, most of these Internet-connected institutions have also adopted the use of leading-edge video conferencing and streaming media technology (such as DVTS or MPEG) to enable students to have access to world-class learning experiences such as music classes taught by world-renowned musicians or scientists anywhere in the world. Students are also encouraged to participate in programs like Megaconference Junior, a project that brings together thousands of students in elementary and secondary schools from around the world using advanced multipoint video conferencing (ibid.).

In addition, not too long ago, students wanting a college diploma or a university degree had to pay tuition and then attend class on a university campus. Today, students wanting to advance their postsecondary education may do so simply with the click of a computer mouse. For example, the Directory of Schools, featured at http://www.directoryofschools.com/, is a Website designed for college and university students seeking online programs through distance learning. This online search engine directory admits students worldwide—from those seeking college diplomas to those wanting doctoral degrees—to universities such as the University of Phoenix and Kaplan University. In all, more than 8,000 degrees and accredited online programs are offered. Although the enlisted colleges and universities would claim to provide an equal educational advantage to citizens around the world—regardless of where they are—the problem arises when experts and law enforcement claim that some of these institutions are merely offering degree-mill diplomas and degrees.

And while some might argue that there is no harm in getting more education, regardless of whether it is received through the Internet or in the traditional classroom, news media reports have in recent years cited some North American tragedies. For example, after a mother named Marion Kolitwenzew discovered that her daughter was diabetic, in 1999 she took her child to a specialist whose walls were filled with a number of alleged medical degrees from seemingly reputable universities, and whose office was brimming with medical supplies. It turned out, however, that this individual's diplomas came from degree mills, defined as bogus postsecondary institutions conferring diplomas and degrees for a fee, often with little or no studying on the part of the registered student. When the mother followed the so-called physician's advice and stopped giving her daughter insulin, the eight-year-old child started vomiting and eventually died. The North Carolina man who treated the child, Laurence Perry, was convicted of manslaughter and practicing medicine without a license. He is currently serving just fifteen months in jail (Armour, 2003).

Other professions have been equally hard hit with false degree holders, including such fields as engineering, sex-abuse counseling, and psychological counseling; even college presidents have appeared with false degrees. The use of diploma mills is exploding as the Internet makes bogus degrees easier to get than ever before. Without question, many more employees around the globe are today purchasing these credentials so that they can get promoted or gain a competitive edge in tough job markets.

Moreover, since traditionally based universities are moving toward more online course offerings, the cyber environment is such that degree mills can flourish more easily than ever before. In fact, as part of their branding exercise, many of these online degree mills intentionally try to mimic the names of well-established postsecondary institutions to trick consumers into registering online. It is interesting to note that in 2002, an investigation by the U.S. General Accounting Office discovered more than

1,200 resumes on a government Internet site with listed degrees from diploma mills. Even the Homeland Security Department's deputy chief information officer, Laura Callahan, was under investigation for her doctoral degree, which apparently was awarded by Hamilton University of Wyoming—not Hamilton College in New York or another legitimate university. Given this modern-day reality, a number of U.S. states have recently passed laws making it a crime punishable by imprisonment to use a bogus degree to get a job (ibid.).

Although citizens around the world tend to think that just about any kind of information is available through the World Wide Web, there remains some information that evades such transmission. For example, most of the books available in libraries and book stores are not available on the Internet, for two main reasons. First, many authors are not willing to give their creative Intellectual Property (IP) to others for free. Second, given the thousands of books available in many libraries today, it would take an extremely long time and huge amounts of money to digitize all the books available through conventional circulation policies (Tardy, 2005).

However, a number of Websites are available to assist online users with personal lifelong learning, including language training. For example, www.Realfrench.net provides a collection of free online French language resources developed by Manchester Metropolitan University (ibid.).

Another reality about Internet access globally is that there is not quite an even playing field in terms of global cultural adoption of and adaptation to the Internet. Particularly in the developing countries, access to the Internet is an issue. It is well known that in China and in India, for example, citizens in the developing world rely on the Internet to obtain Western news and information, but only a portion of the population has Internet access. In China, as a case in point, while there is an estimated population of more than 1.3 billion people, there are an estimated 94 million Internet users—most located in universities or large cities (Thomson, 2006).

Moreover, in the developing world, online access to Western newspapers, such as Canada's *Globe and Mail*, and to Western online reference sources, such as Wikipedia, are often censored by governments, because, they say, they are concerned about the morals of their citizens. In January 2006, for example, communist authorities in China blocked the Wikipedia Website, the third shutdown of the site in China over the past two years. The most recent shutdown continued for ten weeks without any explanation and without any indication of whether the ban would be temporary or permanent (York, 2006b).

There is, unquestionably, a digital divide between those who are wealthy enough to obtain Internet access and those who cannot afford to do so. Moreover, not every government is agreeable to having its citizens roam the Net freely, because, they maintain, there are moral concerns. In such cases, the government feels the need to censor Websites that they feel are harmful to their citizens or to pass laws that make breaches of "community standard" tolerance levels subject to sanctions.

In short, online users in developing nations, in particular, face a series of cultural adaptation and restriction issues that are very complex in nature. These include social, legal, and economical facets. This chapter focuses on some of those critical cultural adaptation and restriction issues, using real-world examples as cases in point.

Cultural Adaptation Factors Influencing the Digital Divide

Geographical Factors

If one considers that the Internet was developed as a security adaptation in the United States (ARPAnet) by the U.S. government so that during times of war, information could be transmitted to critical agents to keep the country secure, then one would expect that the United States would be far ahead in the adoption and adaptation race in terms of Internet access and use for U.S. citizens. Recent Computer Industry Almanac estimates, in fact, confirm that this expectation is, indeed, the case. Although the Almanac projected that worldwide Internet users in 2005 would exceed 1 billion by the end of 2004, as few as fifteen countries accounted for 71 percent of the global Internet user population. As expected, the United States topped the Internet adoption list, accounting for 185 million Internet users. The next heaviest Internet user was China (99 million users), followed by Japan (78 million users), Germany (41 million), India (37 million), the United Kingdom (33 million), South Korea (32 million), Italy (about

26 million), France (about 26 million), Brazil (22 million), Russia (21 million), Canada (20 million), Mexico (13 million), Spain (13 million), and Australia (13 million). In terms of market share for the top three user countries, the United States had 20 percent of the Internet user market, followed by China with 11 percent, followed by Japan with 8 percent (Computer Industry Almanac, Inc., 2004).

The almanac had other predictions about future growth in Internet adoption and adaptation—namely, that there would be little adoption growth in the developed countries, because the market is pretty well at its saturation point, but that over the next five years many present-day Internet users with home PCs would be making use of newer applications, including Voice over Internet Protocol (VoIP) to make less expensive long-distance telephone calls. There was also projected to be an increase in mobile device use among present-day Internet users (with the Research In Motion—RIM—being one such device).

In the developing nations, Internet use was conjectured to continue to climb over the next decade, as more and more citizens gained access. The largest growth was projected to be in China, primarily because of the government's tremendous economic growth. Growth in Internet use was projected to increase but at a slower rate in the developing countries of Pakistan, India, and Russia, where access issues for citizens are more strained (ibid.).

Socioeconomic Status, Gender, and Life Stage Factors

Other demographic variables, including socioeconomic status, gender, and life stage, have been reported to have an impact on Internet-user access and adoption, particularly in the developing nations (Vaagan and Koehler, 2005). For example, Internet access is far greater for citizens in cities than in rural areas—with Internet cafes (that is, storefronts having Internet connections for fee-paying consumers) being quite popular in the more densely populated areas. Moreover, in the developing nations, the well-educated college and university population and those with greater economic means are more likely to have Internet access than are those without advanced education or wealth.

Currently, gender is also a key variable regarding Internet access in the developing world. As in the developed world, men are more likely to gain access to a computer at earlier ages and to be more interested in hacking, cracking, networking, and Internet chat rooms (Schell and Dodge, 2002). In the future, however, as women gain access to the Internet at earlier ages—motivated primarily as a way to communicate inexpensively with their friends—the gender divide may become less acute.

Finally, life stage factors are important in terms of Internet access for those living in developed and developing nations. Individuals born after the 1950s are more likely to seek access to the Internet, because they were more likely to be exposed to computers at younger ages, and are, therefore, less likely to feel threatened by the opportunities that Internet access can provide.

Legal Jurisdictional Adaptation Factors

Clearly, Internet access and adoption by citizens in developed and developing nations can provide for both personal and financial growth. However, Internet access and adoption by the masses can also lead to revengeful online acts or exploits intended to bring personal financial gains to online users maliciously motivated. As these malevolent online exploits have been brought to court, interesting legal jurisdictional questions have arisen. This next section will show how complex legal jurisdictional adaptation factors can become, as more than one jurisdiction becomes involved in any given online exploit.

A U.S. Noncomplex Legal Jurisdiction Case in Point

One such malicious motivation case that made world headlines back in 2001 was that of U.S. waiter Abraham Abdullah. A self-taught cracker, Abdullah was arrested and sent to prison for defrauding financial institutions of about $20,000,000 by using an identity theft scheme. Abdullah would select his targets' identities from the Forbes 400 list of America's wealthiest citizens—including Steven Spielberg, Oprah Winfrey, Martha Stewart, Ross Perot, and Warren Buffett. Then, with the help of his local library's computer, Abdullah would use the Google search engine to glean financial information on these wealthy citizens. He then used information obtained from forged Merrill Lynch and Goldman Sachs correspondence to persuade credit reporting services (such as Equifax and Experion) to supply him with detailed financial reports on those targeted individuals. The detailed reports were then used by Abdullah to dupe banks and financial brokers into transferring money to accounts controlled by him (Caslon Analytics, 2006; Schell and Dodge, 2002).

This case illustrates that with access to the library's computer and the Internet, Abdullah was able to initiate a surprisingly simple process of identity theft by gaining access to credit card and brokerage accounts. Abdullah's scheme came to light when he sent a fake e-mail message to a brokerage house requesting a transfer of $10,000,000 to his bank account in Australia from an account belonging to e-commerce millionaire Thomas Siebel. An American, Abdullah was tried according to U.S. jurisdictional law and was sentenced to imprisonment in a U.S. penitentiary (Caslon Analytics, 2006; Schell and Dodge, 2002).

The Abdullah case also shows the relative simplicity of jurisdictional issues when an online exploit is conducted by a citizen of a given country and no cross-border issues are involved. Internet crimes involving cross-border jurisdictions, however, make for much more difficult prosecution and, often, competing rulings and remedies by the courts. Before discussing some real-world cases, more needs to be said about how nations around the globe generally treat Internet-related crimes.

The Electronic Frontier and Frontier Justice

Without question, monopolies in new industries tend to attract the attention of antitrust laws, and within a short span of time the New Frontier, electronic or not, becomes increasingly disorderly. Governments around the globe have scrambled over the decades to bring order and control to disorderly environments, stretching financial and legal resources and forcing the nation to implement, at times, rather blunt instruments of justice. As a result, the concept of "frontier justice" has become a common term in society's lexicon. Briefly, "frontier justice" encapsulates the struggle by lawmakers and enforcement agencies to bring order and control where there is none. Invariably, affirm lawmakers, "examples" have to be made of offenders. Abraham Abdullah would be one such Electronic Frontier criminal who had to be brought to justice, thus serving as a visible presentation to other online citizens that Internet-related crimes will not be tolerated by U.S. society. Other jurisdictions around the globe have produced similar visible presentations (Schell and Dodge, 2002).

Over time and regardless of jurisdiction, there has been a continual state of disorder-to-order evolution, including that involving the Electronic Frontier. Since the mid-1990s, for example, monthly, it seems, somewhere around the globe, new legislation is being implemented, or a special task force of law enforcement officials is being established, to deal with some evolving facet of Internet crime.

The CAN SPAM Act of 2003—a U.S. act created to thwart online spammers and cyber-pornographers, in particular—serves as one such evolutionary case in point. With the passage of this act, both U.S. citizens and those around the globe appear to have benefited in terms of spam crime reduction. Although almost a quarter of the world's spam in the last quarter of 2005 was sent from U.S.-based computers, according to UK antivirus company Sophos, that is good news. In 2004, about 42 percent of all relayed spam came from U.S.-based computers. The reduction has been linked, say legal authorities, to the effectiveness of the CAN SPAM Act of 2003 and to the U.S. courts' imposition of harsh prison sentences and stiff financial penalties on spam "examples" found guilty of this Internet-related crime.

One such spam case involved the April 8, 2005, sentencing of spammer Jeremy Jaynes of North Carolina. Jaynes received nine years in U.S. federal prison for his online exploits; according to legal authorities, he was labeled one of the Top Ten spammers in the world.

Often, the passage of evolving legislation has drawn criticism from those supporting the values of free speech and free-flowing information in society—a common argument, for example, used by those in the White Hat hacking community and espoused by the Electronic Frontier Foundation (EFF).

These free speech critics have argued that governments around the globe have tended to over-react to a large number of so-called Internet crimes, and, often, they assert, the wrong parties are being made to serve as "examples." White Hat hackers, motivated by a need to find flaws in computer systems and to provide fixes (or patches) for the flaws, are too often labeled as "Black Hat examples" by the news media and legal authorities.

Moreover, the critics affirm, the state's legislative response to threats to computer systems and networks, in particular, is generally skewed by a perceived threat of "outside" crackers, cyber-terrorists, and wrongdoers, when in fact the principal threat of cyber-attacks already exists *within* the country, *within* a business, or *within* the institution that the law seeks to protect from external attack. In the U.S. jurisdiction, both the Abdullah and Jaynes cases would be cited by the critics as two cases in point of "insiders" causing harm as a result of Internet abuses.

In the UK jurisdiction, the London bombings that took place on July 7, 2005, would classify as a noncyber "insider" crime. Those terrorists, who killed fifty-two UK citizens as they were commuting to work, were raised in the Beeston district of Leeds (Suleaman, 2005).

Finally, from a historical perspective, it is generally extremely difficult to assess the "reasonableness" of the laws' response to a perceived threat to societal order and well-being within one's own lifetime and from within one's own culturally biased conceptual and emotional framework. That said, most modern-day legal observers seem to view recently passed Internet crime laws as a "reasonable" compromise of the competing interests espoused by the "property and person paradigm" (that is, proposing remedies for harm done to property or to persons) and by the free speech and free-flowing information contingency (Schell and Dodge, 2002).

One hundred years ago, for example, a judge in the Wild West would not have "reasonably" hesitated to sentence a horse thief to public hanging. Because of the disorder-to-order evolution, however, how the law would "reasonably" treat a horse thief now would be very different from how a horse thief would have been treated earlier: there is little doubt that in modern North American jurisdictions, for example, a horse thief would not be hanged.

Considering the worldwide confusion and early-stage development of Internet crimes, many more controversial laws, court cases, and court renderings are sure to surface around the globe over the next decade. And critics are sure to argue, at times, that too many "examples" have been wrongly charged, or that they have been made to serve too harsh a sentence for a cyber-crime they allegedly committed.

Governments and Internet-Related Crimes

The point has been made that various jurisdictions around the globe have attempted to deal with the evolving facets of Internet-related crime. This section describes some legal attempts made by various countries to cope with this evolution, beginning with the United States and including three cases of violations of the Digital Millennium Copyright Act, then moving to Canada, the United Kingdom, and a more global dimension.

The United States

The Computer Fraud and Abuse Act, the Patriot Act of 2001, and the Digital Millennium Copyright Act (DMCA).

In the United States, the Computer Fraud and Abuse Act (CFAA) was passed in 1986 to deal with computer-related crime, in particular. Through its evolution, this act has been amended in 1994, in 1996, and, again, in 2001 by the USA Patriot Act.

At its inception, the CFAA applied only to government computers. Today, it applies to a much broader group of computers, including the Internet. Furthermore, its purpose has been expanded in a more recent evolutionary twist with the intent of deterring terrorists and cyber-terrorist "examples" from causing harm to property and persons in the United States.

As noted, the Patriot Act of 2001, also known as H.R.3162, was introduced on October 23, 2001, by James Sensenbrenner with the intention of deterring and punishing terrorist acts in the United States and elsewhere and enhancing law enforcement investigation tools. The passing of the amendment was, in part, a reaction to the devastating outcomes of the September 11, 2001, terrorist attacks. Related earlier evolutionary bills included H.R.2975 (an earlier antiterrorism bill that passed the House on October 12, 2001) and H.R.3004 (the Financial Anti-Terrorism Act). On October 26, 2001, H.R.3162 became Public Law No. 107–56, more commonly known as the USA Patriot Act of 2001.

During the 1996 evolutionary phase of the CFAA, the National Information Infrastructure Protection Act (NIIPA) expanded the act to include unauthorized access to and acquisition of information from a protected computer without proper authorization from the party owning that computer.

Before these more recent CFAA amendments, for the U.S. courts to find a violation of the CFAA, the judges would have had to interpret the CFAA as requiring that the accused (that is, "the example") intended commercial gain from the online exploit. The CFAA, said by legal experts to have been drafted with future evolutionary phases in mind, provided the basis for criminal prosecution of cyber-crime in the United States. It was also designed to be easily modified to reflect both changes in technology and changes in criminal techniques.

Currently, the CFAA is broad in its application, reflecting the U.S. government's resolve to combat Internet crime on a comprehensive basis. A conviction for violation of most of the provisions of the CFAA can include up to five years in prison and a fine of up to $250,000 for a first offense. For a second offense, convicted "examples" can serve up to ten years in prison and be made to pay up to a $500,000 fine. The current version of the CFAA also permits any target suffering damages or losses to take civil action against the violator.

It is important to note, as well, that the application of certain subsections of the CFAA to Internet crimes has presented challenges for prosecutors. For instance, the requirement for damages defined to be "a loss aggregating at least $5,000 in value during a one-year period" has been difficult to prove in court, especially when, for example, the text of a Web page has been altered, but the computer system has not been damaged (ibid.).

Another important piece of legislation passed in recent years in the United States to deal with Internet crimes is the Digital Millennium Copyright Act (DMCA). Enacted in 1998, the DMCA was intended to implement under U.S. law certain worldwide copyright laws to cope with emerging digital technologies. By providing protection against the disabling or bypassing of technical measures designed to protect copyright, the DMCA encouraged owners of copyrighted works—such as songs and DVDs—to make them available on the Internet in a "safe" digital format. DMCA sanctions would then apply to anyone trying to impair or disable any encryption device protecting such a copyrighted work.

Jurisdictional Factors in Three Court Cases Involving Violations of the DMCA. Three recent court cases involving violations of the DMCA, in particular, illustrate both how legal jurisdictional issues are critical to the outcomes of those cases and how jurisdic-

tional issues can add to the complexity of the cases' resolution—
especially when more than one jurisdiction is involved.

For example, in 1999, just a year after the passage of the
DMCA, a civil court case arose in the United States that was one
of the more interesting hacker activist cases to make headlines.
Known as the Internet free speech and copyright case to those in
the White Hat hacker community, it involved *2600: The Hacker
Quarterly* and, in opposition, Universal Studios and other members of the Motion Picture Association of America. The case arose
in November 1999, from *2600's* publication of and linkage to a
computer program called DeCSS as part of its news coverage
about DVD decryption software. Universal Studios and fellow
companies filed suit against *2600* in January 2000 and sought a
court order ensuring that the magazine no longer distribute the
computer program in question. These powerful complainants
objected to the publication of DeCSS, because they argued that it
could be used as part of a process to infringe copyright on DVD
movies.

In their defense, the editors of *2600* and representatives from
the EFF argued that decryption of DVD movies is necessary for a
number of reasons, including to make "fair use" of movies and
to be able to play DVD movies on computers running Linux—an
operating system widely used on Internet servers and embraced
by large corporations as an alternative to the Microsoft operating
system software. In the end, the courts agreed with the arguments made by lawyers for Universal Studios and other members of the Motion Picture Association of America. Note how relatively easy this U.S. case was to conclude in a court setting,
primarily because all parties to the complaint were Americans,
and the case arose—and was settled—in the U.S. jurisdiction. In
fact, the resolution of this case was as comparatively noncomplex
as the Abdullah case, described earlier (ibid.).

The second case involving Intellectual Property Rights (IPR)
violations also took place in the United States, but with a for-
eigner as the alleged violator of the DMCA. This case also made
for very interesting reading, but it was a bit more complex than
the former case because two jurisdictions were involved. In July
2001, Dmitry Sklyarov, a Russian scientist who had developed
software to crack Adobe's supposedly secure eBook Reader for
his employer, ElcomSoft, was arrested and prosecuted in Las Vegas as he was about to address the DefCon Computer Underground community. The U.S. government agents who arrested

him argued that Sklyarov was in a U.S. jurisdiction and that his Adobe code-cracking software was in violation of the DMCA. The court case increased in color as Sklyarov's Russian employer was also later charged. The case ended with both Sklyarov and Elcomsoft walking away free. In the end, the U.S. courts concluded that Sklyarov's behavior was legal in the country and in the jurisdiction in which he had developed the software—Russia (Schell and Martin, 2004).

The third case involves a fifteen-year-old Norwegian hacker nicknamed DVD-Jon. This case was very interesting because although the hacker allegedly committed an Internet crime in Norway, charges were brought forward by the U.S. DVD Copy Control Association Incorporated. Again, two jurisdictions were involved, making resolution of the case rather messy, costly, time-consuming, and complex.

The issue started when DVD-Jon developed a software program to circumvent the content scrambling protection on DVDs, making it possible to view DVDs with the open source Linux operating system. He then published the DeCSS decryption program on his Website, and he permitted other Internet users to download the program if they wished. It is important to know that DeCSS removes the copy bar (that is, content scrambling system) in DVDs and stores a copy of the film on the hard disk of the computer.

The issue became an international incident in 2000, because the copy bar is licensed by the DVD Copy Control Association Incorporated for the protection of DVD films produced by Motion Pictures Association members. Thus, in January 2000, the U.S. DVD Copy Control Association (DVD-CCA) and the Norwegian Motion Picture Association (MAP)—affiliated with the U.S. Motion Picture Association of America (MPAA)—filed charges in Norway under penal code section 145. Three years later, DVD-Jon was fully acquitted of all charges in an Oslo municipal court, but the ruling was appealed to the Borgarting appellate court. In the end, the 2003 appellate court ruling upheld the lower court's ruling, and the Norwegian hacker walked away a free man (Vaagan and Koehler, 2005).

The Oslo court found, first, that access to movies legally purchased was not unlawful, even if the movies were viewed differently from the way presumed by the producer of the movie. The Oslo court also found, that disclosure of encryption keys by the minor did not in and of itself constitute unauthorized access to data. In other words, the minor could not be convicted for contributing to unauthorized access to DVD movies by others, because the software program also had a legal application. The appellate court found that the minor's actions were not an infringement of the provisions of the DMCA (ibid.).

The DVD-Jon court rulings were viewed as a clear victory for the White Hat community. The White Hats argued that the freedom of information and the open-source software movement principles were upheld by the Norwegian courts. Equally as important, this case demonstrates the complexity of and alternate findings of courts as the issues in question involve more than one geographical jurisdiction—particularly when they involve different countries.

In combination, these three cases involving the DMCA raise a number of Information Ethics (IE) and legal questions that have yet to be fully answered. Clearly, all three cases are about harm to property, but the Norwegian court rulings also suggest that access and freedom of expression are also in jeopardy. For example, can all kinds of digital information and knowledge be considered to be Intellectual Property—and, if so, is there a clear borderline between public domain software (such as open-source, which is free) and proprietary software (such as Microsoft, which is not free)? Also, how can one distinguish between the public's right to access and the ownership rights of authors and producers with respect to Intellectual Property Rights (Vaagan, 2004)?

Without question, there is a variation in the frequency of DeCSS software Website postings across nations—indicating a more open or a more closed orientation to this software's presentation to citizens. Also, there is a variation in the context in which DeCSS is shown on Websites—indicating either a likely negative software application or a likely positive software application.

For instance, some nations (such as The Netherlands, Germany, the United Kingdom, France, and the United States) contain more examples of DeCSS Website postings than do other nations (such as Italy, Spain, Ireland, and China). Moreover, there are differences in the context created by Website authors for DeCSS. Although authors in China tend to present DeCSS as an audio-visual cracking tool (with placement on Web page categories such as "DVD Rippers" and "Decoding"), most European Union authors tend to present DeCSS in a more politically open context or without any clear positive or negative context (with placement on Web page categories such as "Linux DVD" and "DeCSS Central"). Although sociopolitical stances as well as global differences in law provide some explanation for these observed variances, there are many issues left unanswered at this stage of Internet crime evolution. And there are many variations in interpretation from one jurisdiction to another (Eschenfelder, 2005).

Canada

Compared with the United States, Canada's attempts at dealing with Internet crime have been more compartmentalized, with prohibitions found in the Canadian Criminal Code. Moreover, an Internet offense generally falls in one of two categories, depending on the severity of the crime: an indictable offense, or a summary conviction offense.

Indictable offenses are more serious than summary conviction offenses; therefore, the former carry lengthier maximum prison sentences. In combating Internet crimes in Canada, the following criminal code provisions are generally applied: theft, fraud, computer abuse, data abuse, and the interception of communications.

The primary cracking prohibition in the criminal code is set out in Section 342.1—namely, "unauthorized use of a computer," often referred to as the "computer abuse" offense. Section 342.1 of the criminal code is aimed at several potential harms: protection against theft of computer services (342.1(1)(a)), protection of privacy (342.1(1)(b)), and protection against persons who trade in computer passwords or who crack encrypted systems (342.1(1)(c)). Of all of these provisions, section (c) is likely the most controversial, because it creates as an offense the mere use of a computer to *intend* to commit offenses (a) or (b), or to commit mischief pertaining to data. The underlying rationale for provision (c) is that law enforcement and potential targets of Internet crime should not be required to wait until actual harm to property or persons is inflicted before the act of abuse attracts criminal sanction—a point that has angered White Hat hackers, who feel that many in the Computer Underground can be unjustly criminalized.

By now, most readers are probably familiar with the case of a young Canadian cracker named Mafiaboy. He is the fifteen-year-old minor who in February 2000 raised concerns in North

America and elsewhere about Internet safety when he cracked Internet servers and used them as launching pads for Denial of Service (DoS) attacks on the Websites of Amazon, eBay, and Yahoo. He admitted his part in the cyber-attacks in January 2001, in a courtroom in Montreal, Quebec. The judge ruled that the minor had indeed committed a criminal offense in violation of the criminal code, and he was sentenced to eight months in a youth detention center. Mafiaboy was placed on one year's probation after his release and was fined $250. Like the Abdullah case in the United States, this Canadian case is not that legally complex, because the Canadian youth was convicted in a Canadian jurisdiction and served time in a Canadian detention center.

Other well-publicized, Internet-related crime cases have also had interesting legal nuances regarding cross-border jurisdictional issues involving Canada and the United States. One such case was *Guillot v. Istek Corporation* (2001), a copyright-infringement claim over online articles and links posted on a Website. The plaintiff was a U.S.-based trademark lawyer who maintained that a Canadian Website copied both articles and a series of links posted on his Website without getting authorization from him. The trademark lawyer asked the courts for an injunction forcing the Canadian Website to immediately remove the disputed content. However, the judge hearing the case refused to issue the injunction; he found that posting material on the Internet includes an implied license to copy material to the extent necessary to make personal use of it. There was no evidence in this case that the Canadian Website did anything that it was not implicitly authorized to do. Michael Geist, a Canadian professor of Internet law, said that while the judge's decision highlights the challenge of applying conventional copyright principles to the Internet, it also demonstrates the difficulty of effectively explaining rapidly emerging technologies to those unfamiliar with the most recent Internet developments (Bareskin and Parr, 2005; Geist, 2001).

The United Kingdom

The main Internet crime law passed in the United Kingdom within the recent past is the Computer Misuse Act of 1990, brought in after the failed prosecution of Robert Schifreen and Steven Gold—the famous 1986 crackers of the BT Prestel text information retrieval system. Although the pair were convicted on a number of charges under the British Forgery and Counterfeiting Act of 1981, on appeal they were acquitted on the grounds that the language of the British Forgery and Counterfeiting Act was not intended to apply to this case. Therefore, the Computer Misuse Act was passed to prevent unauthorized access to computer systems, to deter criminals from using computers to commit property crimes, and to prevent criminals from impairing or hindering access to data stored on a computer.

More recently, the United Kingdom passed the Terrorism Act of 2000 to criminalize public computer cracks, particularly when the crack puts the life, health, or safety of UK persons at risk. The United Kingdom, like other jurisdictions with serious economic interests in the Internet—including the United States and Canada—has chosen to adopt an approach to Internet abuse legislation that results in criminal sanctions by linking cracking activities to matters of fundamental national interest.

International Developments: The Convention on Cyber-Crime

In 2001, group action on an international level was being taken to combat Internet crime. On November 23, the Council of Europe opened to sign its newly drafted Convention on Cybercrime. The convention was signed by thirty-three states after the council recognized that many Internet crimes could not be prosecuted with existing legislation—typically local in jurisdiction. The convention was the first global legislative attempt to set standards on the definition of Internet-related crime and to develop policies and procedures to govern international cooperation to fight it.

The treaty was to enter into force when five states, at least three of which were members of the Council of Europe, had ratified it. The United States, as a participant in the drafting of the treaty, had been invited to ratify the treaty. In many adopting states, ratification of the treaty would require amendments to national law. It is interesting to note that President Bush transmitted the convention to the U.S. Senate on November 17, 2003, for ratification. Finally, the convention was adopted at the 110th session of the Committee of Ministers in Vilnius, on May 3, 2002.

The convention requires countries ratifying it to adopt similar criminal laws on cracking, infringements of Intellectual-Property Rights, Internet-related fraud, and Internet-related child pornography. It also contains provisions on investigative powers and procedures, including the search of computer networks and the interception of communications. In particular, the convention requires cross-border law enforcement cooperation in searches and seizures as well as extradition. The convention has recently been supplemented by an additional protocol making any publication of racist propaganda via the Internet a criminal offense (Center for Democracy and Technology, 2005).

Culture and the Digital Divide

So far, this chapter has discussed some key access issues regarding the Internet. First, there was a discussion about how various geographical, socioeconomic status, gender, and life stage factors influence global citizens' Internet access. Next, there was a discussion about how different nations around the globe have dealt with unauthorized computer access and Internet misuse breaches. This section will present an international incident that triggered Internet community standard tolerance breach questions, caused deaths, evoked cultural restriction reactions from various nations around the globe, and produced news media and legal questions about whether it was, indeed, a case of cyber-terrorism.

An International Incident

In late December 2005, a young Dane by the name of Ahmed Akkari flew to Beirut, carrying a package of booklets in green covers containing cartoons depicting the Prophet Mohammed. A young Islamic scholar and activist, he was on a mission. He failed to get the Danish prime minister to take action over the cartoons' perceived slight to Islam when they were published in Danish newspapers and posted on Websites. Akkari was equally disturbed about the slight to the Prophet Mohammed brought about by the degrading nature of the media- and Internet-published cartoons (Saunders, 2006a).

In a matter of weeks, Akkari had given copies of the booklets to the grand mufti of Egypt, the chief cleric of the Sunni faith, leaders of the Arab League, and the head of the Lebanese Christian Church, to name just a few. Their reaction was quite predictable: they stared in amazement at the degrading images of the Prophet Mohammed. As of February 8, 2006, Ahmed Akkari found himself doubting whether his actions had done more harm than good. He knew that as a result of his actions, millions of online users saw somewhat distorted cartoons flash around the Muslim world through the Internet, through newspaper photos, and through verbal text-message descriptions. Violent protests erupted about the cartoons and continued to erupt around the globe, bringing the death toll from the demonstrations to at least nine. The United Nations, the European Union, and other governments struggled to contain the escalating violence and unrest (ibid.).

During the first week of February 2006, the Norwegian and Danish embassies in Damascus, Syria, were set on by angry Muslims, and on February 5, 2006, the Danish consulate in Beirut was also burned to the ground. Ironically, the anger sweeping the globe was not started with the Internet images or the text messages; it began with a children's book. In 2005, Danish writer Kare Bluitgen was searching for someone to illustrate his account of the Prophet Mohammed, but he could not find anyone to do the job, primarily because individuals feared antagonizing Muslims' feelings about images of their prophet. By disclosing his dilemma to several Danish newspapers, newspaper editors began to question how far Denmark should go in self-censorship— and whether freedom of speech was more important than Muslims' cultural sensitivities (McKay, 2006).

Then, on September 30, 2005, the editor of the Danish newspaper *Jyllands-Posten* launched a provocative experiment. He published images from twelve cartoonists whom he had hired to create satirical drawings of the Mohammed. One photo, for example, depicted a bearded Mohammed with his turban transformed into a bomb. Another showed a group of ragged suicide bombers arriving in heaven, only for the prophet to tell them, "Stop, stop, we've run out of virgins." Another showed Mohammed as a dog engaged in offensive actions. Within days, three of the cartoonists who had contributed to the newspaper's display received death threats, and security guards were placed outside the newspaper's offices in Copenhagen and Arhus. What should have remained a local controversy became an international incident within a few days after Prime Minister Anders Fogh Rasmussen and the Danish judicial system failed to act or to do damage control (ibid.).

Moreover, as part of the globe exploded in furious and, at times, deadly riots against the Danish cartoons, the streets of Copenhagen suddenly took on an air of frightened calm, as the majority of Danish citizens feared that the cartoon furor would lead to terrorist attacks. Although some of the Danish citizens also felt insulted by the cartoons because they believed that the depictions surpassed their community standard tolerance levels, others felt that the global protests were unfair slurs on the country's much-cherished culture of free expression and freedom of speech. Without doubt, the majority of Danish citizens were concerned about the harm to the economy that began to quickly manifest. For example, Iran announced that it was severing all trade ties with Denmark as part of its promise to respond to an anti-Islamic current, and Cairo protesters called for an immediate boycotting of all Danish and Norwegian products (Saunders, 2006b).

International Reactions to the Incident. The U.S. reaction to the Islamic terrorist attacks was restrictive, calling for the Danes and other countries around the globe to snoop on Islamic telephone conversations and to read Islamic text messages. Attorney General Alberto Gonzales affirmed on February 7, 2006, that snooping on telephone calls is vital to winning the war against Islamic terrorists; he in fact suggested that wiretaps imposed earlier by President George W. Bush had thwarted terrorist strikes in the United States after 9/11. The secret surveillance program, he said, began after the September 11, 2001, air attacks on the World Trade Center and on the Pentagon (Koring, 2006).

Elsewhere, on February 7, 2006, French president Jacques Chirac pledged solidarity with Denmark. Also, the European Commission in Brussels released a statement saying that no real or perceived grievance justifies acts of violence (Sachs, 2006).

However, the European leaders—unable in the recent past to create a common foreign or defense policy as they had been able to create a common foreign cyber-crime policy—appeared to be unwilling and unable to forge a uniform response to the newly erupted violence. Moreover, the attacks on European offices in the Gaza Strip, in the first week of February 2006, also raised concerns that the cartoon incident could poison future relations with Palestinians, already unsettled by a recent election resulting in a victory for Hamas, the Islamic fundamentalist group (ibid.).

Rather than calling for a restriction of or an intervention in Islamic communications, Canadian foreign affairs minister Peter MacKay noted on February 8, 2006, that while he condemned the violence and simultaneously defended world citizens' freedom of expression, he affirmed that this right to freedom of expression must be exercised responsibly. This sensitive issue, he said, highlights the need for a better understanding of Islam and of Muslim communities by citizens around the globe. He affirmed that the Canadian government pledges to promote a better understanding of Islam internationally, while partnering with Muslim communities. On an aside, while some Canadian university newspapers printed the cartoons, none of the mainstream newspapers did (Sallot and Den Tandt, 2006).

Meanwhile, on February 8, 2006, the Canadian government was, nevertheless, worried about the safety of more than 2,000 Canadians being deployed to Afghanistan, a Muslim country with some of the bloodiest of the anti-Western protests, following the cartoon's release. For example, on February 8, Afghan police killed four protestors, bringing the number of cartoon-related deaths in that country to eleven (ibid.).

Elsewhere on February 8, 2006, crackers defaced as many as 600 Danish Websites, replacing the usual information with threats and hacktivist messages. One page said: "Danish, you'r D3ad," while another showed a mannequin painted in the Danish flag and hanging from its neck. On that day, Muslims in Bangladesh burned Italy's flag and called for the boycott of European Union goods (ibid.).

The Turks, angry that the leaders of Denmark and other European countries had apologized for insulting Islam but not for publishing the caricatures (repeatedly in their newspapers and online), took part on February 7 and 8, 2006, in demonstrations outside the Danish consulate and the French cultural center. As the furor over the cartoons continued to grow, some Turks began to question whether there was room for a country of 72 million Muslims in a secular but predominantly Christian European Union (MacKinnon, 2006).

Moreover, on February 9, 2006, advice given to hundreds of business leaders, economists, and institutional investors around the world by experts at U.S.-based Goldman Sachs was blunt and went something like this: Given the terrorist atmosphere that seems to predominate around the globe, build your bomb shelters and buy a bullet-proof vest (Grant, 2006).

Legal Issues Raised by This Incident.   This case raises two very interesting legal questions that deal with restrictions in the cyber world. In terms of the community standard, did the newspaper editor exceed the community's normative standard by initially publishing these cartoons in hard copy and then online—thus breaching Danish law, followed by likely breaches in other jurisdictions around the globe? Also, since this case involved the Internet, elements of terrorism (that is, the burning of buildings), online death threats to citizens, and resulting deaths from demonstrations, is this case, by definition, a case of cyber-terrorism?

The Community Standards Tolerance Level Issue Raised by This Incident.   Without question, in the last fifteen years, there has been a meteoric rise in the popularity of a new medium of communications and entertainment called the Internet. This new medium has not only brought the promise of new opportunities for self-expression and the communication of ideas to a broad, global audience; it has also brought the opportunity for breaches in levels of tolerance within communities (since anyone with an Internet account can post information on the Net and do so relatively uncensored).

In fact, when an online user posts information on the Net, that person is known to be a "content provider." As the Internet has grown in popularity with the growth of home PCs, so have citizens' concerns around the globe about objectionable or obscene material available on Websites that can be accessed by minors—such as written and graphic pornography (particularly child pornography), hate literature, and downloadable information on bomb-making and computer-cracking (Sillito, 1999).

What is obscene material? The U.S. court system, for example, tends to utilize the Miller test for determining whether free speech or free expression is deemed to be "obscene" and, therefore, not protected by the First Amendment. If information meets this test, it can legally be banned. The Miller test stemmed from a court case known as *Miller v. California*, a 1973 court case in which the U.S. Supreme Court held that material is found to be obscene if these factors are satisfied (EFF, 2006):

- Whether the average person, by applying contemporary community standards, would perceive that the work in question and taken as a whole appeals to the prurient interest;
- Whether the work in question depicts or describes in a patently offensive way sexual conduct defined by applicable laws; and
- Whether the work taken as a whole lacks serious literary, artistic, political, or scientific value.

Countries around the globe have developed similar tests for determining if information exceeds that jurisdiction's community standards. In most developed countries, for example, pornography depicting sexual acts among consenting adults or genitalia would not be considered to be obscene, but in developing nations, community standards tend to vary widely, and those nations tend to be more conservative in their definitions.

Under current international laws, the legal question of whether Internet-related speech is deemed to be obscene is determined, in part, by reference to a local community's tolerance standards. Federal venue rules tend to allow an obscenity prosecution to be brought where the online speech originated, or where it was received. Since Internet speech is able to be received in every community and in all nations, the community standards criterion as applied to a nationwide audience tends to be judged by the standards of the community most likely to be offended by the online message (ibid.).

Even within countries, there can be differences in community standard tolerance levels, as the latter vary from one jurisdiction to another (Schell and Bonin, 1988, 1989; Schell and Bigelow, 1990). As a case in point, in the *U.S. v. Thomas* case, a U.S. Appellate Court in the 1990s held that a couple who operated a computer Bulletin Board System (BBS) in California could successfully be prosecuted for breaching the Digital Millennium Copyright Act (DMCA) if the case were to be held in a more conservative jurisdiction.

The latter case depended on the community standard definition of obscenity. Because the community standard tolerance level in western Tennessee was far more conservative than that found in California, the prosecutors for the case thought that Tennessee would be a better venue for hearing the case than California would be. Although the accused couple argued for protection of freedom of speech under the U.S. First Amendment, the pair was eventually convicted under a number of federal statues, including 18 U.S.C. §1465. They were found to have *intentionally* used a facility and a means of interstate commerce—in this case, a computer and the telephone system—to transport obscene, computer-generated information. In short, the pair ran an Internet pornographic BBS from their California home by scanning onto their BBS sexually explicit photos that they bought in various adult book stores. To view and download these photos, Internet users had to pay a $55 membership fee to the pair and sign an application that included spaces for users' names and addresses (Rogers and Cohen, 1998).

On an international perspective, the publishing of the cartoons of the Prophet Mohammed can arguably be seen as a breach of community standards across global jurisdictions, particularly those containing a majority of Muslim citizens. Although the violent outcomes that ensued are unjustified and even criminal, perhaps they were effective in having culturally insensitive countries realize their wrongdoings, particularly with regard to breaches of international community-standard tolerance levels.

The United Kingdom seems to have acted in a culturally sensitive way in the days following the Danish publication of the offensive cartoons. The British airwaves tried to strike a balance between the various holders of community-standard tolerance levels and to elicit mutual respect in this instantly connected Internet world by inviting representatives advocating free speech and those calling for the cartoonists to be tried in an Islamic court to discuss their thoughts on the air so that all citizens could hear.

There was also a riveting discussion broadcast on BBC's *Newsnight* program that allowed two British Muslim women to calmly argue against violence with extremist Anjem Choudary of the al-Guraba group, who called for violence against those favoring the offensive cartoons. It is clear that the British saw the cartoon eruption not as a terrorist war but as a cultural-sensitivity and community-standard breach. Viewed in that way, it becomes clear that this cartoon dispute will not be won by snooping on Muslim's cellular phones but by permitting public and respectful communication exchanges. Offering platforms of civilized free speech for European Muslims to debate with one another in an appropriate context—as Britain did—benefits society much more than heavy-handed restrictions or violence (Ash, 2006).

The Incident as an Act of Cyber-Terrorism. Regarding the question of whether the cartoon online publication actually turned into an act of cyber-terrorism because it involved the Internet, elements of terrorism (that is, the burning of the buildings), and resulting deaths of protestors, it is clear that there is no easy or obvious answer to this question, because of the very early evolution of this legal term.

In the present, the meaning of the word *cyber-terrorism* is being debated by Internet researchers, government and legal policy makers, scholars, and Internet security firms. The word was coined in the 1980s by Barry Collin, a senior research fellow at California's Institute for Security and Intelligence. It was derived by joining the concepts of "cyberspace" and "terrorism" (a term believed to be derived from the Latin word *terrere*, meaning "to tremble" or "to frighten").

The intrinsic problem is that the word *terrorism* itself has very different meanings for different people. In short, because the Internet is still relatively young, the meanings of words used to describe breaches are still evolving, both in definition and in legal circles around the globe. Moreover, depending on one's vantage point, parties to the debate of definition have taken sides over whether cyber-terrorism as a crime is an immediate problem, a future threat, or an overhyped concept used by the news media and governments. Finally, because of legal and cultural differences among nations, it is difficult to respond effectively to online attacks. It is doubly difficult to deal with Internet breaches when there is disagreement about both the definition of what constitutes cyber-terrorism and the degree of danger to society likely to result (Keith, 2005).

In an attempt to provide scholars with a framework for understanding how the word *cyber-terrorism* has been used by the news media in recent years, in 2005, Susan Keith released her findings about how the term had been used in 146 articles appearing between 2001 and 2004 in mainstream English-language U.S., British, and Canadian newspapers and magazines. She then produced the Cyber-terrorism Coverage Model, which had variables relating to the notions that cyber-terrorism is alive and real, that fear among government agents and citizens is expected and justified, that defense and protection against the crime are critical, and that cyber-security experts are needed to stop cyber-criminals in their tracks. Keith also had variables in her model relating to the overhyped news media angle. On the one hand, there was the notion that cyber-terrorism is more hype than reality; on the other hand, there was the notion that present-day terrorists lack the needed technical skills to cause real harm to society, that present-day terrorists perceive the impact of cyber-terrorism to be less effective than that of old-fashioned bomb attacks in targeted "soft" areas (such as trains and subways), and that public awareness about cyber-terrorism is important but that civil liberties should be preserved (ibid.).

Indeed, Keith's work sets out nicely how the news media in various jurisdictions have in recent times conceptualized for citizens how the term *cyber-terrorism* may be perceived and acted upon by citizens and various agencies. However, until the courts around the globe actually rule on real cases and attempt to answer the legal questions raised here and found in Keith's model, for the time being the Prophet Mohammed cartoon case cannot be clearly pronounced as being a case of "cyber-terrorism."

Cultural Property and Personal Access

The final section of this chapter deals with cultural-property restriction and personal-access restriction issues. The first part describes key legal concepts around Intellectual Property Rights and agencies created to protect such rights. The second part of this section describes how different nations, particularly developing nations, have moved to restrict citizens' access to certain Websites because they are believed to be harmful to the moral values of that society. In short, the final portion of this chapter focuses on Internet censorship in certain parts of the globe.

Intellectual Property
Rights (IPR) Protection

There are a number of laws and protection agencies around the globe to protect Intellectual Property Rights.

The Role of WIPO. One such agency is the World Intellectual Property Organization (WIPO), an international organization promoting the use and protection of Intellectual Property—the creative outputs in all fields, including science, technology, and the arts—and available for posting online.

With its head office in Geneva, Switzerland, WIPO is one of the sixteen specialized agencies of the UN system of organizations. WIPO's primary function is to administer twenty-three international treaties dealing with different aspects of IP protection, and it currently has 183 nations as members (all of which appear on WIPO's Website; WIPO, 2006).

Before the advent of the Internet, IP was and continues to be protected by patents, copyright, and trade secrets. But with advances in technology and with the worldwide growth of the Internet, other legal and technological measures have been created to deal with IP abuses. For example, anticopying restriction measures such as copy bars are often used by proprietary software producers to curb piracy and illicit copying. As a legal example, the Digital Millennium Copyright Act (DMCA) was passed in the United States in 1998 to provide protections against technical measures that could be utilized to disable or bypass encryption devices used to protect copyright. The DMCA was created to encourage authors of copyrighted material to place their work on the Internet in a digitalized presentation.

Moreover, penalties are to be applied when IPR breaches occur. For example, DMCA penalties are to be applied to any individual attempting to disable or successfully disabling an encryption device. Stated simply, according to this piece of legislation, Intellectual Property and copyright infringement are theft—taking what does not belong to the perpetrator of the encryption bypass, thereby depriving the true copyright owners of royalties for the sale of their creative products.

Global Concerns about Piracy. It is interesting to note that although 183 nations are part of WIPO, legal allegations have arisen in recent years that in some countries, piracy is nonetheless openly justified or covertly condoned by some of the member nations. This kind of behavior is not acceptable, say member nations who uphold the tenets of the agreement. As a result, some WIPO nations have argued that unauthorized access restrictions need to be espoused by and reinforced through sanctions by WIPO participants. On the contrary, arguments have been made—particularly by the developing nations—that poverty prevents poorer citizens in their countries from making legal purchases of software or DVDs. Therefore, pirating or reverse-engineering imported materials represents redress for decades of exploitation of the poorer class (Vaagan and Koehler, 2005).

A Recent Case of IPR Infringement. To illustrate a recent case of IPR infringement, on May 22, 2005, reports surfaced that counterfeiters in Beijing, China, were selling illegally copied DVDs of the film *Star Wars: Episode III-Revenge of the Sith* just days after the film opened in theaters in North America. The price charged for the pirated movies, sold from vendors wearing shoulder bags on the streets of Beijing, was a mere $3.05. The street sales occurred, despite numerous Chinese government promises to clamp down on the thriving black-market industry that North American movie companies have argued costs them billions of dollars in lost revenue yearly. About 9,000 cases of piracy were brought to court in China in 2004 (Associated Press, 2005).

Is there factual evidence to corroborate the claims that pirated goods are a potential means of redressing decades of exploitation of the poorer class, particularly in developing countries? Although this question likely cannot be answered here, the reality is that the poorest 20 percent of China's urban residents are earning less than 3 percent of the total urban income, according to a new study released by China's National Development and Reform Commission. Moreover, the inequality is even more obvious when assets are measured. The wealthiest 10 percent of Chinese urban residents control a significant 45 percent of urban assets, while the poorest 10 percent of residents have only 1.4 percent of the assets. The sad news is that not only has the earnings-and-assets gap reached an alarming level in China but the gap appears to be expanding. Although the Chinese government says that one of its 2006 objectives is to narrow the income gap, a number of sociologists around the globe maintain that the huge gap will continue to cause hatred of the rich by the poor for quite some time (York, 2006c).

Relative to China, recent economic reports indicate that India fares far worse. Some economic experts say that this is the price that India is paying for its democracy. The economic reforms in India started in 1991 with Prime Minister Narasimah Rao, who released the country from the suffocating web of bureaucratic socialism. Despite the wealthier software and business services area present in Bangalore, the rest of India is still struggling economically—with the majority of citizens surviving at near the poverty level. In short, India is no China. Whereas China's powerful government officials can bulldoze slums to build modern-day highways and increase the country's wealth potential, India's leaders find themselves tirelessly negotiating with their own citizens to make any real economic progress, or just to stay economically functional. As a case in point, the Indian government finally reached settlement with its airport workers following a five-day strike in 2006 and hours upon hours of negotiation. During negotiations, the area surrounding the airport smelled bad—a problem for potential business investors and visitors—the result of the hot temperatures fermenting heaps of stinking garbage (Mortished, 2006).

European Efforts to Protect Intellectual Property Rights. Elsewhere around the globe, as a means to further reinforce the protection of IPR in Europe, in April 2004 the European Union Directive for the Enforcement of Intellectual Property Rights became law. It is important to point out that no matter where in the world IPR legislation is passed, there is always controversy. With the passage of this directive, for example, came complaints from the Irish Software Organization that the law's extreme provisions and its tough sanctions toward regular consumers for accidental infringements appear to be unreasonable. Moreover, because this directive had tough sanctions for those wanting to reverse-engineer licensed software products to produce competing but lower-priced items, open-source and free software developers were concerned that their products could be sanctioned (Vaagan and Koehler, 2005).

Censorship of Certain Websites

Censorship by some nations of certain Websites is a current social global issue. Censorship of information is one social issue that has plagued public librarians for years when they select material for their library collections. Although researchers have reported that the censorious librarian tends to be female, to have less than a university education, and to have generally liberal attitudes but conservative, risk-averse behaviors (Schell and Bonin, 1988), there has been little research conducted by scholars to shed light on why certain countries—such as China and Russia—feel the need to censor their citizens' access to certain Websites.

Possible Explanations for Governments' Censorious Behavior. As the Internet has grown to become an ingrained part of Internet-connected citizens' lives, a number of concerns about commercial harms have been raised. Concerns about the darker side of the Internet have given governments in both developed and developing nations a reason for investigating a number of Websites deemed potentially harmful to users, particularly to minors. Targeted Websites for investigation by government and legal authorities have included those created for often controversial commercial purposes—such as pornography distribution, interactive gaming, online auctions, and online gambling.

Moreover, concerted efforts to prevent harm to persons that may be caused by the viewing of and interaction on certain commercial Websites and chat rooms have come from a number of sources, including online industries, consumer interest groups, legislators, and government agencies. Claims of potentially harmful consequences to online citizens have, in fact, led to numerous calls for the regulation of the contents of commercial Websites through a number of measures. In the United States, for example, restrictive regulations have included the passage of the Child Online Protection Act (COPA). This piece of legislation was created to restrict obscene materials from being available to minors online. Also, the Project Safe Bid, to curb online auction fraud, was created, as was the Internet Gambling Prohibition Act to ban online gambling.

The growing willingness to restrict commercial Websites by governments and citizens appears to rest on the belief that such restrictions will help reduce online pornography distribution, online violations of IPR and consumer privacy, as well as online auction fraud and online gambling addictions. Proponents of such restrictive measures have argued that legal interventions can protect vulnerable groups—children, teenagers, and women

—from the dangerous materials illustrated on or the unsafe activities promoted by the Internet.

However, whenever legal measures are created to restrict access, controversy typically ensues. One piece of legislation resulting in controversy in the United States, for example, was the Communication Decency Act, or CDA. As a means of restricting minors' access to indecent and patently offensive speech on the Internet, in 1996 the U.S. Congress passed the CDA. However, shortly after its passage, a lawsuit was launched by the American Civil Liberties Union, alleging that this piece of legislation violated the First Amendment. The U.S. Supreme Court, supporting that view, struck down the CDA. A more recent and second attempt to regulate pornography on the Internet resulted in the passage of COPA. By remedying the alleged defects in the CDA, COPA was made to apply only to those communications made for commercial purposes and considered to be potentially harmful to teens or children.

It is important to note, however, that the debate concerning the regulation of Website content and the protection of minors has not been limited to pornographic commercial Websites. Websites promoting online games have also provoked a number of voiced concerns by parents, educators, and politicians about the potential harm they can cause, including that of motivating online users to act out in real-life aggressive acts. Given that children or teens in developed countries today have nearly unlimited access to the Internet, it is plausible, they have argued, that without proper monitoring or parental restrictions, minors can easily access game sites promoting interactive violence. In fact, some online gaming critics have blamed violent computer games for desensitizing young gamers to the ugliness of bloodshed and homicide. As a response to public outrage about online gaming, politicians in many jurisdictions have provided resources to help parents protect their children from objectionable or offensive material posted on the Internet.

Online auction sites are also major points of Internet controversy. According to e-commerce analysts at Forrester Marketing Research, the number of products sold online increases dramatically year after year. In 1998, for example, only about $1.4 billion worth of products were sold through online auctions, but by 2003, online auction sales rose to more than $19 billion. Although online auction sales continued to boom in 2006, consumer protection groups have increasingly advocated for restrictions to be placed on online auction Websites because of an increased incidence of online fraud and the sale of controversial products. For example, in 1999 a human kidney was placed for sale on the eBay Website, receiving bids by online clients of nearly $6 million (Yeon, 2001).

In the United States, the Federal Trade Commission (FTC), concerned that there needed to be a vehicle for alerting prospective online auctioneers that fraud was becoming more prevalent online, launched Project Safebid—a full-blown effort to get law enforcement agents to treat online auction fraud cases very seriously and to apply strong sanctions for those who defraud others. Today, auction Websites tend to cooperate with law enforcement agents to prevent the selling of offensive items online—including drugs, weapons, and body parts, human or otherwise. Often, caught and convicted spammers and online-fraud perpetrators are sent to jail (ibid.).

Censorship Cases in China. North America is not the only place where Internet controversies and restrictions have occurred. In China, for example, the government tends to monitor what users do and see on the Internet. The Chinese government has invented two cartoon figures, called Jingjing and Chacha, that float on users' computer screens. Their job is to keep online users under constant surveillance to ensure that they do not view material perceived by the government to be unauthorized or subversive. Users, for example, who try to post comments on any of Shenzhen's Websites or online chat rooms will find that the cartoon characters pop up on their computer screens to remind users that their online behavior is being monitored. The line used by the Chinese government is that the cartoons' official function is to maintain order on the Electronic Frontier and to remind online citizens to be conscious of the safe and healthy use of the Net. What is very interesting is that the cartoon figures were designed by real-life Internet police in Shenzhen, China, so that Internet users could communicate interactively with the cartoon police (and, thereby, real police). The combined names of Jingjing and Chacha produce the word "jingcha," which means "police" in Chinese. Censored Websites by the Chinese government in 2006 have included the Canadian mainstream newspaper *The Globe and Mail*, blocked and then unblocked a number of times. With an estimated 40,000 Internet police, China is actively engaged in a campaign to control what online users see and read (York, 2006a).

Although censorship of Chinese Websites is seen as a necessary restrictive action by the government to keep citizens safe, apparently not all citizens are happy with the restrictions. Chinese students and intellectuals have openly voiced outrage at Beijing's decision to restrict access to the Wikipedia Website, an online encyclopedia that has become a resource for many students and academics in China and elsewhere. The site offers more than 2 million articles in 100 languages, but its presentation on political issues such as the Tibetan autonomy issue, Taiwanese independence, and the Tiananmen Square student protests has evoked the wrath of the Chinese government. As a reaction to one of the blockages, one message posted on an online site said, "What idiots these officials are! They are killing our culture with censorship" (York, 2006b).

Reactions to Chinese Censorship by Business. Although China wants to do business with the rest of the world—online or in the conventional sense—censorship like this can create havoc for Chinese businesses connected to the World Wide Web. For example, in August 2005, North America heard about the newly released Chinese search engine Baidu.com when its stock soared more than 350 percent on the first day of trading. The thirty-seven-year-old Chinese founder, Robin Li, told newspaper reporters that although Chinese Internet usage was still in its infancy and was constantly being monitored by the government, he, like North American search engine founders, had hopes of becoming another high-tech billionaire, because Chinese citizens enjoy staying "connected" with the rest of the world through the World Wide Web. Although Li's statement contained much truth, on October 5, 2005, the reality was that Chinese authorities shut down another two Internet sites popular among academics, journalists, and civil rights activists. There appear to be no guarantees that Li's Website may not some day face a similar fate.

Today, businesses wanting to do transactions in China are making adaptations. For example, in February 2006, the Microsoft Corporation outlined a policy for handling government restrictions on personal Websites. General counsel Brad Smith said that the software maker would shut down personal Web logs, or blogs, only if it received a legally binding notice from a government. In the past, Microsoft did not require that content explicitly violate laws for it to be blocked or removed. Mr. Smith affirmed that the company would block a site deemed to be illegal only in the country where it received legal notice—but nowhere else. The company wanted to maintain access to the site for the rest of the noncensored world. In a similar vein, Google said that its new Chinese service would remove links to Websites believed to be offensive to the Chinese government. In particular, Google affirmed that they would not run certain services—like blogs and e-mail—on computers in China (Guth, 2006).

Censorship Cases in Russia.   China is not the only country that wants to keep a careful eye on its citizens' online activity. In Russia, twenty-six-year-old Andrei Skovorodnikov, a young guitarist, has just finished serving a six-month conditional sentence for publishing an offensive phrase on the Internet—or, as the indictment noted, for repeating the words twenty-five times in Russian and English. The phrase included these words: "Oh, oh, oh, Putin. Oh, oooh, Putin is a fag." Although Andrei says that his Website was meant to be a joke, he did not think it was a joke when the police raided his apartment, threw him into prison for almost two years, and placed his name in the criminal files. He was convicted under Article 319 of Russia's criminal code, forbidding public insult of an authority representative, connected with the fulfillment of his responsibilities. It is interesting to note that Reporters Without Borders ranks Russia near the bottom of its press-freedom index: 140th among 167 countries (Smith, 2006). In comparison, China ranks 159th (York, 2006a).

Conclusion

This chapter has discussed the cultural, economic, legal, and political impact of the Internet in developed countries such as the United States, Canada, and the United Kingdom, and developing countries like China, India, and Russia. The focus throughout the chapter was on critical cultural adaptation and restriction issues related to the growth of the Internet around the globe.

Compared with the development of the Internet in North America—where companies like Yahoo, Inc., have been able to succeed because they got onto the Information Game highway early (thanks to the trail-blazing efforts of ARPAnet), with plenty of financing, with the ability to hire the best professionals to make the business initiative a success, and with a minimal degree of restricted user access—by this chapter's end it should be clear that in China, as a comparative case in point, Baidu, the Chinese Internet search engine that finally emerged as a growing positive concern in news media articles in the summer of 2005, had been five years in the making. The odd thing is that, like Yahoo, Inc., and ARPAnet, Baidu also has a U.S. connection.

Although relatively unknown outside of China until now, Baidu was founded by two Chinese gurus who had previously worked for U.S. technology firms. The entrepreneurs who own Baidu claim that they have a strong following of more than 100 million online Web surfers in China, but they openly admit that, relative to the open Internet surfing privileges given North American and European online users, the Chinese government censors whatever content it feels might corrupt the morals of its citizens.

In short, this chapter has focused on the unique cultural adaptations and restrictions regarding the Internet found in the underdeveloped and developing nations around the globe. The strong message has been that the Internet is still in its early phases of evolution; many more cultural adaptations, legislative challenges, and restrictive interventions are likely to arise in the forthcoming years.

References

Armour, Stephanie. "Diploma Mills Insert Degree of Fraud into Job Market." USA Today, http://www.usatoday.com/money/workplace/2003–09–28-fakedegrees_x.htm (accessed September 29, 2003).

Ash, Timothy. "Going Head to Head for Islam's Heart." *Globe and Mail*, February 9, 2006.

Associated Press. "Entertainment: Counterfeiters Move Fast on Illegal Star Wars DVD." *Globe and Mail*, May 23, 2005, p. B7.

Bareskin and Parr. "Recent Court Decisions 2001." Bareskin and Parr, http://www.bareskinparr.com/English/publications/rcd/pub_rcd_aug_2001.html (accessed July 4, 2005).

Caslon Analytics. "Caslon Analytics Profile: Identity Theft, Identity Fraud." Caslon Analytics, http://www.caslon.com.au/idtheftprofile2.htm (accessed March 2006).

Center for Democracy and Technology. "International Issues: Cybercrime." Center for Democracy and Technology, http://www.cdt.org/international/cybercrime/ (accessed February 8, 2005).

Computer Industry Almanac, Inc. "Worldwide Internet Users Will Top One Billion in 2005." Computer Industry Almanac, http://www.c-i-a.com/pr0904.htm (accessed September 3, 2004).

EFF (Electronic Frontier Foundation). "Bloggers FAQ-Adult Material." Electronic Frontier Foundation, http://www.eff.org/bloggers/lg/faq-adult.php (accessed February 22, 2006).

Eschenfelder, Kristin. "Chasing Down the Social Meaning of DeCSS: Investigating the Internet Posting of DVD Circumvention Software." American Society for Information Science and Technology, http://www.asis.org/Bulletin/Jun05/eschenfelder.html (accessed June/July 2005).

Geist, Michael. "Cyberlaw Shows Its True Colours." *Globe and Mail*, September 6, 2001, p. B27.

Grant, Tavia. "Terrorism and Oil Woes Top List of Biggest Financial Worries." *Globe and Mail*, February 9, 2006, p. B19.

Guth, Robert. "Microsoft Revises Policy on Spiking Blogs." *Globe and Mail*, February 1, 2006, p. B11.

Keith, Susan. "Fear-Mongering or Fact: The Construction of 'Cyber-Terrorism' in U.S., U.K. and Canadian News Media." Oxford Internet Institute, http://www.oii.ox.ac.uk/research/cybersafety/extensions/pdfs/papers/susan_keith.pdf (accessed September 8–10, 2005).

Koring, Paul. "Phone Snooping Needed, Gonzales Says." *Globe and Mail*, February 7, 2006, p. A13.

MacKinnon, Mark. "Looking West, but Still Deeply Offended." *Globe and Mail*, February 9, 2006, p. A17.

Magill, K. "Merchants—Not Consumers—Are Blocking E-commerce, Study Says." *Direct Marketing News* 20 (1998): 19–20.

McCarthy, J., with M. Tavilla. "Business View and IT Brief." http://www.forrester.com/findresearch/results?SortType=Date&Ntt=%246.8+trillion&swin=1&Ntk=MainSearch&Ntx=mode+matchallany&geo=0&dAg=10000&N=50181 (accessed March 13, 2001).

McKay, Simeon. "Text Messages Helped Stoke Flames of Fury." *Globe and Mail*, February 6, 2006, pp. A1, A12.

Mortished, Carl. "India Booms but—Unlike China—Democracy Dulls the Tiger's Claws." *Globe and Mail*, February 9, 2006, p. B17.

Nowak, G., et al. 1999. "Interactive Media: A Means for More Meaningful Advertising?" In *Advertising and the World Wide Web*, edited by D. Schumann and E. Thorson, 99–117. Mahwah, NJ: Lawrence Erlbaum Associates, Inc.

Rogers, S., and B. Cohen. "The First Amendment and the Internet: United States v. Thomas, 74 F. 3d 701 6th Cir. 1996." Sugarman et al., http://www.netlitigation.com/netlitigation/cases/thomas.htm (accessed 1998).

Rotman, Lauren. "Internet Survey Finds over 46,000 K20 Institutions Connected to Its Next-generation Network." Lauren Rotman, https://mail.internet2.edu/wws/arc/i2-news/2006–03/msg00004 .html (accessed March 24, 2006).

Sachs, Susan. "Europe Fails to Mount United Response to Protests." *Globe and Mail*, February 7, 2006, p. A13.

Sallot, J., and M. Den Tandt. "Cartoons Caused Offence, Ottawa Says." *Globe and Mail*, February 9, 2006, p. A17.

Saunders, Doug. "'It Is Not What I Want to Happen.'" *Globe and Mail*, February 8, 2006a, pp. A1, A14.

_____. "Danes Fear Cartoon Furor Will Lead to Terror Attacks." *Globe and Mail*, February 7, 2006b, pp. A1, A13.

Schell, B., and B. Bigelow. "Expert Testimony in Pornography Trials: A Role for the Social Scientist." *Behavioral Sciences and the Law* 8 (1990): 301–311.

Schell, B., and L. Bonin. 1989. "Understanding Pornographic Tolerance Levels of Community Residents Regarding Three Media: Magazines, Movies, and Video Cassettes." *Annals of Sex Research* 1 (1989): 501–521.

_____. 1988. "Factors Affecting Censorship by Canadian Librarians." *Journal of Psychology* 123 (1988): 357–367.

Schell, B., and J. Dodge, with S. Moutsatsos. 2002. *The Hacking of America: Who's Doing It, Why, and How*. Westport, CT: Quorum.

Schell, B., and C. Martin. *Cybercrime: A Reference Handbook*. Santa Barbara, CA: ABC-CLIO, 2004.

Sillito, Casey. "The Relationship between Attitude toward Government Censorship of the Internet and Internet Exposure." http://s93502979 .onlinehome.us/cjsweb/employ/portfolio/academic/psych_4.htm (accessed February 25, 1999).

Smith, Graeme. "'It Wasn't That Funny the Day They Raided My Apartment.'" *Globe and Mail*, January 31, 2006, p. A3.

Suleaman, Nasreen. "The Mystery of Sid." http://news.bbc.co.uk/1/ hi/magazine/4354858.stm (accessed October 19, 2005).

Tardy, Yvan. "Internet for Learning Languages." University of Bristol, http://www.vts.rdn.ac.uk/tutorial/langs?sid=7425100&op=render&m anifestid=132&page=copyrighthtml (accessed April 2005).

Thomson, Sandy. "China Rising." *Elle Canada* 59 (May 2006): 116–118, 120.

Vaagan, Robert. "Ethical Issues Raised by the Electronic Information Environment." *Proceedings of the Annual Conference of the Library Association of Ireland and the Chartered Institute of Library and Informational Professionals (CILIP) Ireland,* 2004.

Vaagan, R., and W. Koehler. "Intellectual Property Rights vs. Public Access Rights: Ethical Aspects of the DeCSS Decryption Program." Information Research, http://informationr.net/ir/10–3/paper230.html (accessed April 2005).

WIPO. "About WIPO." World Intellectual Property Organization, http://www.wipo.int/about-wipo/en/ (accessed February 9, 2006).

Yeon, Seounmi. "Censorship of Commercial Websites." Michigan State University. http://list.msu.edu/cgi-bin/wa?A2=ind0109b&L=aejmc &T=0&F=&S=&P=3185 (accessed September 9, 2001).

York, Geoffrey. "Chinese Ban on Wikipedia Prevents Research, Users Say." *Globe and Mail,* January 10, 2006a, p. A2.

_____. "Internet Police State." *Globe and Mail,* January 28, 2006b, p. F1.

_____. "China Frets over Its Expanding Income Gap." *Globe and Mail,* February 9, 2006c, pp. A1, A18.

4

Chronology

The Setting

1940s and 1950s In the 1940s and 1950s, computers are made with 10,000 vacuum tubes and occupy more than 93 square meters of space. There is a limit to how big computers can be because they could overheat and explode. Major improvements come in computer hardware technology with the development of transistors in 1947 and 1948. Computers develop even more with the advent of integrated circuits in 1958 and 1959.

1960s Since the 1960s, the number of transistors per unit area has been doubling every one-and-a-half years, increasing computing power tremendously. This amazing progression of circuit fabrication is called Moore's Law.

1968 The Intel company in the United States is started by Andy Grove, Gordon Moore, and Robert Noyce to build semiconductor memory products. Their 2006 company Website speaks to their huge success over the years:

> "This year 100 million people around the world will discover digital for the first time. This year 150 million more people will become part of the wireless world. This year the living room will grow more interactive and the digital divide will shrink. This year more people will be using technology in more fascinating ways than ever imagined. And behind all of this progress you'll find innovative Intel technology."

1969 ARPAnet (Advanced Research Projects Agency Network) starts, the first intercontinental, high-speed computer network built by the U.S. Department of Defense as a digital communications experiment. By linking hundreds of universities, defense contractors, and research laboratories, ARPAnet allows Artificial Intelligence researchers in dispersed areas to exchange information with incredible speed and flexibility. This capability advances the field of Information Technology (IT). Instead of working in isolated pockets, the White Hats are now able to communicate via the electronic highway as networked tribes—a phenomenon still existing in today's Computer Underground.

The standard operating system UNIX is developed by Bell Laboratory researchers Dennis Ritchie and Ken Thompson. UNIX is considered to be a thing of beauty because its standard user and programming interface assist users with computing, word processing, and networking.

The Creative Beginnings: The 1970s

1970 An estimated 100,000 computer systems are in use in the United States.

1971 Canadian Stephen Cook publishes Cook's Theorem, which helps to advance the field of cryptography.

1972 The National Center for Supercomputing Applications creates the telnet application for remote login, providing an easier means for users to log into a remote machine.

David Boggs and Robert Metcalfe invent Ethernet at the Xerox Corporation in California.

1973 Intel's chairman, Gordon Moore, publicly reveals the prediction that the number of transistors on a microchip would double every year and a half. This prediction is known as Moore's Law.

The File Transfer Protocol (FTP) is developed, simplifying the transfer of data between networked machines.

Canadian Mers Kutt creates Micro Computer Machines and releases the world's first Personal Computer (PC).

1975   Apple Computer, Inc., is founded by a pair of members of California's Homebrew Computer Club: Steve Jobs and Steve Wozniak. Once the Apple Computer and the simplistic BASIC language appear on the hacking scene, techies see the potential for using microcomputers.

William Henry Gates, III (commonly known as "Bill Gates"), and Paul Allen form Microsoft, Inc.

1976   The Diffie-Hellman Public-Key Algorithm, or DH, is developed by Whitfield Diffie and Martin Hellman. The DH, an algorithm used in many secure protocols on the Internet, is now celebrating more than thirty years of use.

1978   By the end of the 1970s, the only positive thing missing from the cyber community is a form of networking social club. In 1978, that void is filled by two men from Chicago, Randy Suess and Ward Christiensen, who create the first computer Bulletin Board System (BBS) for communicating with others in the wired world of computers.

The Transmission Control Protocol (TCP) is split into TCP and IP (Internet Protocol) by Vinton Cerf.

The Growing-up Years: The 1980s

1981   IBM (International Business Machines) announces a new model, stand-alone computer, dubbed "the PC."

The "Commie 64" (more conservatively labeled the Commodore 64) and the "Trash-S" (more conservatively

1981 labeled the TRS-80) become two of the enthusiasts' fa-
cont. vorite tech-toys.

1982 A group of talented UNIX hackers from Stanford University and the University of California at Berkeley create Sun Microsystems Incorporated on the foundation
that UNIX running on relatively cheap hardware would
prove to be a perfect solution for a wide range of applications. These visionaries turn out to be right. The Sun
Microsystem Network increasingly replaces older computer systems like the VAX and other time-sharing systems in corporations and in universities across North
America. In 2005, the company Website indicates that,
from a financial perspective, Sun Microsystem Networks
ends the fiscal year with a cash and marketable debt securities balance of more than U.S.$75 billion.

Scott Fahlman types the first online smiley :-).

The Internet is formed when ARPAnet splits into military and civilian sections.

The SMTP (Simple Mail Transfer Protocol) is published.

William Gibson coins the term *cyberspace*.

1983 The Telnet protocol is published.

1984 Steven Levy's book *Hackers: Heroes of the Computer Revolution* is released, detailing the White Hat Hacker Ethic—
a guiding source for those in the Computer Underground to this day.

Fred Cohen introduces the term *computer virus*.

2600: The Hacker Quarterly is founded by Eric Corley
(a.k.a. Emmanuel Goldstein).

Cisco Systems, Inc., is started by a small number of scientists at Stanford University. Even today, the company
remains committed to developing Internet Protocol–

based networking technologies, particularly in the areas of routing and switches.

Richard Stallman begins constructing a clone of UNIX, written in "C" and obtainable to the wired world for free. His project, called the GNU (which means that Gnu's Not Unix) operating system, becomes a major focus for creative hackers.

In Montreal, Gilles Brassard and Charles Bennett release an academic paper detailing how quantum physics can be used to create unbreakable codes using quantum cryptography.

1985   The hacker "zine" *Phrack* is first published by Craig Neidorf (a.k.a. Knight Lightning) and Randy Tischler (a.k.a. Taran King).

www.Symbolics.com is assigned, becoming the first registered domain.

America Online (AOL) is incorporated under the original name of Quantum Computer Services.

The Free Software Foundation (FSF) is founded by Richard Stallman. FSF is committed to giving computer users the permission to utilize, study, copy, change, and redistribute computer programs. The FSF not only promotes the development and use of free software but also helps to enhance awareness about the ethical and political issues associated with the use of free software.

1986   In Britain, the term *criminal hacker* is first alluded to and triggers the public's fears in April 1986 with the convictions of Robert Schifreen and Steven Gold. Schifreen and Gold crack a text information retrieval system operated by BT Prestel and leave a greeting for his Royal Highness the Duke of Edinburgh on his BT Prestel mailbox. The two are convicted on a number of criminal charges under the Forgery and Counterfeiting Act of 1981.

1986 cont.
The Internet Engineering Task Force (IETF) is formed to act as a technical coordination forum for those who work on ARPAnet, on the U.S. Defense Data Network (DDN), and on the Internet core gateway system.

The U.S. Congress brings in the Computer Fraud and Abuse Act. This act is amended in 1994, 1996, and in 2001 by the USA Patriot Act. This act in all of its variations is meant to counteract fraud and associated activity aimed at or completed with computers.

1988
Robert Schifreen's and Steven Gold's convictions are set aside through appeal to the House of Lords, because, it is argued, the Forgery and Counterfeiting Act of 1981 is being extended beyond its appropriate boundaries.

Robert Morris, Jr., becomes known to the world when, as a graduate student at Cornell University, he accidentally unleashes an Internet worm that he has developed. The worm, later known as "the Morris worm," infects and subsequently crashes thousands of computers. Morris receives a sentence of three years' probation, 400 hours of community service, and a $10,500.00 fine.

Kevin Mitnick secretly monitors the e-mail of both MCI and DEC security officials. For these exploits he is convicted of damaging computers and stealing software and is sentenced to one year in prison—a cracking-then-prison story that is to repeat over the next few years.

The Computer Emergency Response Team (CERT)/CERT Coordination Center for Internet security is founded in 1988, in large part as a reaction to the Morris worm incident. Located at Carnegie Mellon University, the center's function is to coordinate communications among experts during security emergencies.

The U.S. Secret Service secretly videotapes the Summer-Con hacker convention, suspecting that not all hacker activities are "White Hat" in nature.

1989   A group of West German hackers led by Karl Koch (affil-
        iated with the Chaos Computer Club) are involved in the
        first cyber-espionage case to make international news
        when they are arrested for cracking the U.S. govern-
        ment's computers and for selling operating-system
        source code to the Soviet KGB (the agency responsible
        for State Security).

        Herbert Zinn (a.k.a. Shadowhawk) is the first minor to
        be convicted for violating the Computer Fraud and
        Abuse Act of 1986. Zinn cracks the American Telephone
        & Telegraph (AT&T) computer systems and the Depart-
        ment of Defense systems. He apparently destroys files
        estimated to be worth about $174,000, copies programs
        estimated to be worth millions of dollars, and publishes
        passwords and instructions on how to exploit computer
        security systems. At age sixteen, he is sent to prison for
        nine months and is fined $10,000.

The Booming 1990s

1990   At the Cern laboratory in Geneva, Switzerland, Tim
        Berners-Lee and Robert Cailliau develop the protocols
        that become the World Wide Web (WWW).

        ARPAnet (Advanced Research Projects Agency Net-
        work) ceases to exist.

Early   There is Internet growth on a homeowner scale, because
1990s   more tech-savvy computer individuals can afford to
        have machines at home that are similar in power and
        storage capacity to the systems of a decade earlier—
        thanks to newer, lower-cost, and higher-performing PCs
        with chips from the Intel 386 family. The down side is
        that affordable software is still not available.

1991   Linus Torvald initiates the development of a free UNIX
        version for PCs using the Free Software Foundation's
        toolkit. His rapid success attracts many Internet hackers
        —who give him their feedback on how to improve his

1991   product. Eventually Linux is developed, a complete
cont.   UNIX built from free and redistributable sources.

The PGP (Pretty Good Privacy) encryption program is
released by Philip Zimmermann. Later, Zimmermann
becomes involved in a three-year criminal investigation,
because the U.S. government says that the PGP program
is in violation of export restrictions for cryptographic
software.

Until 1991, the Internet is restricted to linking the military and educational institutions in the United States. In
this year, the ban preventing Internet access for businesses is lifted.

1992   The Michelangelo virus attracts a lot of news media attention because, according to computer security expert
John McAfee, it is believed to cause great damage to data
and computers around the world. Those fears turn out to
be greatly exaggerated, as the Michelangelo virus actually does little to the computers it invades.

The phrase *surfing the Net* is coined by Jean Armour Polly.

1993   Scott Chasin starts BUGTRAQ, a full-disclosure mailing
list dedicated to issues about computer security, including vulnerabilities, methods of exploitation, and fixes for
vulnerabilities. The mailing list is today managed by
Symantec Security Response.

Just slightly more than 100 Websites exist on the Internet,
and the first Def Con hacker convention occurs in Las
Vegas.

Randal Schwartz uses the software program "Crack" at
Intel for what he thinks is appropriate use for cracking
password files at work, an exploit for which he later is
found guilty of illegal cracking under an Oregon computer crime law.

Linux competes on reliability and stability with other commercial versions of UNIX, and it hosts vastly more "free" software.

1994   News media headlines are sizzling with the story of a gang of crackers led by Vladimir Levin. The gang cracks Citibank's computers and makes transfers from customers' accounts without authorization, totaling more than $10 million. Although in time Citibank recovers all but about $400,000 of the illegally transferred funds, that positive end to the story is not featured by the news media. Levin gets a three-year prison sentence for his exploits.

Canadian James Gosling heads a creative team at Sun Microsystems with the objective of developing a programming language that changes the simplistic, one-dimensional nature of the Web. That feat is accomplished, and the name given to the programming language is Java.

Two Stanford University students, David Filo and Jerry Yang, start their cyber-guide in a campus trailer as a way of tracking their interests on the Internet. The cyber-guide later becomes the popular www.Yahoo.com (which means "Yet Another Hierarchical Officious Oracle").

1995   White Hat hacktivists squash the Clipper proposal, one that would have put strong encryption (the process of scrambling data into something that is seemingly unintelligible) under U.S. government control.

Linux becomes stable and reliable enough to be used for many commercial applications.

Randal Schwartz, writer of the hot-selling books *Programming Perl* and *Learning Perl,* is convicted on charges of industrial espionage. While employed at Intel as a system administrator, he performs security tests using a program called "Crack" to uncover weak passwords. Schwartz is sentenced to five years' probation and almost 500 hours of community work, and has to pay Intel almost $70,000 in restitution.

1995   CyberAngels is created in the United States. It is the
cont.   world's oldest and largest online safety organization.
          Their mission is the tracking of cyber-stalkers, cyber-ha-
          rassers, and cyber–child pornographers.

          The Apache Software Foundation, a nonprofit corpora-
          tion, evolves after the Apache Group convenes in 1995.
          The Apache Software Foundation eventually develops
          the now-popular Apache HTTP Server—which runs on
          virtually all major operating systems.

          The first online bookstore, www.Amazon.com, is
          launched by Jeffrey Bezos.

          Tatu Ylonen releases the first SSH (Secure SHell) login
          program, a protocol for secure remote logins and other
          secure network services over a network previously
          deemed to be nonsecure.

          Microsoft releases Windows 95 and sells more than a
          million copies in less than five days.

1996   Kevin Mitnick is arrested once more for the theft of
          20,000 credit card numbers, and he pleads guilty to ille-
          gal use of stolen cellular telephones. His status as a re-
          peat cyber-offender earns him the cute nickname "the
          lost boy of cyberspace." Computer security consultant
          Tsutomu Shimomura, in close association with *New York
          Times* reporter John Markoff, helps the FBI to eventually
          locate Mitnick, while on the lam. Shimomura and
          Markoff write a book about the episode, calling it, *Take-
          down: The Pursuit and Capture of Kevin Mitnick, America's
          Most Wanted Computer Outlaw—By the Man Who Did It*.
          The book infuriates many in the hacker community.

          White Hat hacktivists mobilize a broad coalition not
          only to defeat the U.S. government's rather misnamed
          "Communications Decency Act" but also to prevent cen-
          sorship of the Internet.

          The National Information Infrastructure Protection Act
          of 1996 (NIIPA) is enacted in the United States to amend

the Computer Fraud and Abuse Act (CFAA), originally enacted in 1984.

The Child Pornography Prevention Act (CPPA) is passed in the United States to curb the creation and distribution of child pornography, including online.

The Internet has more than 16 million hosts.

1997   ARIN, a for-profit organization, assigns Internet Protocol (IP) address space for North America, South America, sub-Saharan Africa, and the Caribbean. Networks allocated before 1997 are recorded in the ARIN whois database.

The DVD (Digital Versatile Disc) format is released, and DVD players are released for sale.

1998   The central activities of the White Hat hacker labs become Linux development and the mainstreaming of the Internet. Many of the gifted White Hats launch Internet Service Providers (ISPs), selling or giving online access to many around the world—and creating some of the world's wealthiest corporate leaders and stock options owners.

The Digital Millennium Copyright Act of 1998 (DCMA) is passed in the United Stated to implement certain worldwide copyright laws to cope with emerging digital technologies. By providing protection against the disabling or bypassing of technical measures designed to protect copyright, the DMCA encourages owners of copyrighted works to make them available on the Internet in digital format.

Cryptographic products from the United States intended for general use outside the U.S. can not legally use more than 40-bit symmetric encryption and 512-bit asymmetric encryption. The reason for this restriction is that the 40-bit key size is widely recognized to be not secure.

1998
cont.
Members of the Boston hacker group L0pht testify before the U.S. Senate about vulnerabilities associated with the Internet.

Canadian Tim Bray helps to create an index known as Extensible Markup Language, or XML, which makes possible the popular online auction eBay.com.

1999
A grand jury in Virginia indicts Eric Burns, aged nineteen years, on three counts of computer intrusion. Burns's moniker on the Internet is Zyklon, and he is believed to be a group member claiming responsibility for attacks on the White House and Senate Websites during this time. A woman named Crystal, who is a cyberstalked target and classmate of his, identifies Burns as Zyklon to the FBI. The judge hearing his case rules that Burns should serve fifteen months in federal prison, pay $36,240 in restitution, and not be allowed to touch a computer for three years after his prison release.

The Internet is infected by the Melissa virus. It moves rapidly throughout computer systems in the United States and Europe. In the United States alone, the virus infects more than 1 million computers in 20 percent of the country's largest corporations. Months later, David Smith pleads guilty to creating the Melissa virus, named after a Florida stripper. The virus is said to cause more than $80 million in damages to computers worldwide.

The Gramm-Leach-Bliley Act of 1999 is passed in the United States to provide limited privacy protections against the sale of individuals' private financial information. The intent of the act is to stop regulations preventing the merger of financial institutions and insurance companies. However, by removing these regulations, experts become concerned about the increased risks associated with financial institutions having unrestricted access to large databases of individuals' personal information.

The Napster music file–sharing system, often used by individuals to copy and to swap songs for free, gains popularity on Websites where students have access to high-speed Internet connections. Napster, developed by university students Shawn Fanning and Sean Parker, attracts more than 85 million registered users before it is shut down in July 2001 as a violator of the Digital Millennium Copyright Act (DCMA).

Jon Johansen, aged fifteen, becomes one of a triad of founders of MoRE (which stands for "Masters of Reverse Engineering"). Johansen starts a flurry of negative activity in the DVD marketplace when he releases DeCSS, a software tool used to circumvent the Content Scrambling System (CSS) encryption protecting DVD movies from being copied illegally.

2000   Authorities in Norway raid Jon Johansen's house and take his computer equipment. Although he is charged with infringing Intellectual Property Rights (IPR), he is eventually acquitted by the courts. His nickname in papers is DVD-Jon.

Another newsworthy hacktivist case is the Internet free speech episode of *2600: The Hacker Quarterly*. For Emmanuel Goldstein, the magazine's editor, the "enemy" is Universal Studios and other members of the Motion Picture Association of America. The civil court legal issue revolves around the DeCSS DVD decryption software and the coverage that Emmanuel Goldstein gave to it in *The Hacker Quarterly*. In the end, the civil court battle favors Universal Studios and the Digital Millennium Copyright Act.

The high-profile case of a Canadian cracker with the moniker Mafiaboy (his identity is not disclosed because he is only fifteen years old at the time) raises concerns in North America about Internet security following a series of Denial of Service (DoS) attacks on several high-profile Websites, including Amazon, eBay, and Yahoo! On January 18, 2001, Mafiaboy says he is guilty of cracking Internet servers and using them to start DoS attacks. In September 2001 he is sentenced to eight months in a youth prison and is fined $250.

2000   John Serabian, the CIA's information issue manager, says
cont.   in written testimony to the U.S. Joint Economic Committee that the CIA is detecting with increasing frequency
the appearance of government-sponsored cyber-warfare
programs in other countries.

IBM estimates that online retailers could lose $10,000 or
more in sales per minute if service is not available to customers because of Denial of Service attacks.

The Love Bug virus is sent from the Philippines. Michael
Buen and Onel de Guzman are suspected of writing and
distributing the virus.

Microsoft Corporation admits that its corporate network
has been cracked and that the source code for future
Windows products has been seen. The cracker is suspected to be from Russia.

In excess of 55,000 credit card numbers are taken from
Creditcards.com, a company that processes credit transactions for e-businesses (that is, those online). Almost
half of those stolen credit card numbers are publicized on
the Internet when an extortion payment is not delivered.

Fear of a High-Tech Meltdown:
2001 to the Present

2001   In a piece published in *The New Yorker,* Peter G. Neu-
mann, a principal scientist at the technological consulting firm SRI International and a consultant to the U.S.
Navy, Harvard University, and the National Security
Agency (NSA), underscores his concerns about the ad-
verse impact of cyber-criminals. He says that he is worried about an imminent cyber-apocalypse, because malicious hackers can now get into important systems in
minutes or seconds and can wipe out one-third of the
computer drives in the United States in a single day.

The Code Red worm compromises several hundred thousand systems worldwide in less than fourteen hours, overloading the Internet's capacity and costing about $2.6 billion worldwide. It strikes again in August 2001. Carolyn Meinel, an author of a number of hacking books and a contributor to *Scientific American*, labels the worm a type of computer disease that has computer security researchers more worried than ever about the integrity of the Internet and the likelihood of imminent cyber-terrorist attacks. She likens the Code Red worm to electronic snakebites that infect Microsoft Internet Information Servers—the lifeline to many of the most popular Websites around the world.

Russian Dmitry Sklyarov is arrested at the DefCon 9 hacker convention in Las Vegas when he is about to give a speech on software particulars that he developed for his Russian employer, ElcomSoft Co., Ltd. The software in question permits users to download e-books from secure Adobe software to more commonly used PDF computer files. The San Francisco–based advocacy group Electronic Frontier Foundation (EFF) lobbies heavily against his conviction, saying that jurisdictional issues apply and that his behavior is perfectly legal in the country in which he performs his exploits (Russia).

The Anna Kournikova virus is placed on the Internet by Jan de Wit (a.k.a. OnTheFly), aged twenty, who is from The Netherlands. He is later arrested and made to perform 150 hours of community service for his exploits.

The *Los Angeles Times* reports that crackers have attacked a computer system, controlling the distribution of electricity in California's power grid for more than two weeks and causing a power crisis. According to the newspaper, the attack appears to have originated from individuals associated with China's Guangdong province. The cyber-attack, routed through China Telecom, adversely affects California's leading electric power grid and causes a lot of concern among state and federal bureaucrats about the potential for a cyber-apocalypse.

2001
cont.

NIMDA (ADMIN spelled backward) arrives, a blend of computer worm and virus. It lasts for days and attacks an estimated 86,000 computers. NIMDA demonstrates that some of the cyber-weapons available to organized and technically savvy cyber-criminals today have the capability to learn and adapt to their local cyber-environments.

On September 11, the United States and the world forget the fears of the Cold War and come face to face with fears surrounding terrorism and cyber-terrorism when Al-Qaeda terrorists deliberately crash passenger jets into the twin towers of the World Trade Center (WTC) and the Pentagon.

On October 23, the USA Patriot Act of 2001 is introduced by James Sensenbrenner, with the intent of deterring and punishing terrorist acts in the United States and of enhancing law enforcement investigatory tools. The introduction of this act is a reaction to the September 11, 2001, terrorist attacks. Related bills include an earlier antiterrorism bill, which passes the House on October 12, 2001, and the Financial Anti-Terrorism Act.

By October 26, just three days after the USA Patriot Act is introduced, it becomes law. Immediately after its passage, controversy is widespread. For example, Congressman Ron Paul informs the *Washington Times* that no one in Congress was permitted to read the Patriot Act beforehand.

Apple Computer releases the iPod portable music player, considered by many to be one example of a good hack that can utilize the song downloading potential of the Internet.

Also, online gaming is becoming a positive social force for the Internet. Online gaming, or Massively Multi-player Online Role-Playing Game (MMORPG), is introduced, a form of computer entertainment played by one or more individuals connected to the Internet.

On November 23, the Council of Europe opens to sign its newly drafted Convention on Cybercrime. The convention is signed by thirty-three states after the council recognizes that many cyber-crimes cannot be prosecuted by existing laws, or that applying those existing laws to cyber-crimes means stretching the intent of the laws. The convention is the first global legislative attempt of its kind to set standards on the definition of cyber-crime and to develop policies and procedures to govern international cooperation to combat cyber-crime.

2002   On July 10 and 11, a U.S. bill on Homeland Security is introduced by Richard Armey to the Standing Committees in the House. The bill is received in the Senate on November 19, 2002, and is passed by the Senate on November 25, 2002. The Homeland Security Act of 2002 is signed by the president as Public Law 107–296 and is meant to establish the Department of Homeland Security. Section 225 is known as the Cyber Security Enhancement Act of 2002.

A seventeen-year-old female cracker from Belgium, known as Gigabyte, claims to have written the first-ever virus in the programming language C# (that is, C sharp).

A fifty-two-year-old Taiwanese woman named Lisa Chen pleads guilty of pirating hundreds of thousands of software copies worth more than $75 million. The software was apparently smuggled from Taiwan. She is sentenced to nine years in a U.S. prison, one of the most severe sentences ever given for such a crime.

2003   A Texas jury acquits a computer security analyst by the name of Stefan Puffer, who a year earlier had been charged with illegally accessing the county computer network. After he had figured out that the Harris County district clerk's wireless computer network was vulnerable, he warned the clerk's office that anyone with a wireless network card could gain access to their sensitive data.

2003
cont.

On April 30, some particulars around the definition of child pornography change when George W. Bush signs the PROTECT Act. This act not only implements the Amber alert communication system—which allows for nationwide alerts when children go missing or are kidnapped—but also redefines child pornography to include both images of real children engaging in sexually explicit conduct with adults and computer images indistinguishable from real children engaging in such acts. Prior to the enactment of the PROTECT Act, the definition of child pornography came from the 1996 Child Pornography Prevention Act (CPPA). The longer name of the PROTECT Act is Prosecutorial Remedies and Tools against the Exploitation of Children Today Act of 2002.

Paul Henry, vice president of CyberGuard Corporation, an Internet security firm in Florida, says that experts predict that there is an 80 percent probability that a cyber-attack against critical infrastructures in the United States could occur within two years. The capability is present among certain crackers and terrorists, Henry warns. It is simply a question, he affirms, of the intent of such criminals to launch an attack.

In July, a poll of more than 1,000 U.S. adults by the Pew Internet and American Life Project finds that one in two adults has expressed concern about the vulnerability of the national infrastructure to terrorist attackers. The poll finds that 58 percent of the women polled and 47 percent of the men fear an imminent attack. More than 70 percent of the respondents are optimistic, however, for they are fairly confident that the U.S. government will provide them with sufficient information in the event of another terrorist attack—be it on the ground or through cyberspace.

Sean Gorman of George Mason University makes headlines when he produces for his doctoral dissertation a set of charts detailing the communication networks binding the United States. Using mathematical formulas, Gorman had probed for critical infrastructure links in an attempt to respond to the query, "If I were Osama bin Laden, where would I want to attack?"

In August, three crippling worms and viruses cause considerable cyber damage and increase the stress levels of business leaders and citizens about a possible cyber-apocalypse. The Blaster worm surfaces on August 11, exploiting security holes found in Microsoft Windows XP. The Welchia worm surfaces on August 11, targeting active computers. It goes to Microsoft's Website, downloads a program that fixes the Windows holes (known as a "do-gooder"), and then deletes itself. The most damaging of the three irritants is the e-mail–borne SoBigF virus, the fifth variant of a "bug" that had initially invaded computers in January and resurfaces with a vengeance on August 18, 2003. The damage for lost production and economic losses caused by these worms and viruses is reportedly in excess of $2 billion for just an eight-day period.

John McAfee, the developer of the McAfee antivirus software company, claims that there are more than 58,000 virus threats. Also, the antivirus software company Symantec further estimates that ten to fifteen new viruses are discovered daily.

On August 14, fears of a cyber-apocalypse heighten for a period known as the Blackout of 2003. The East Coast of the United States and the province of Ontario, Canada, are hit by a massive power blackout—the biggest ever affecting the United States. Some utility control-system experts say that the two events—the August computer worm invasions and the blackout—might have been linked, because the Blaster worm, in particular, may have degraded the performance of several lines connecting critical data centers used by utility companies to control the power grid.

On September 15, the Department of Homeland Security, along with Carnegie Mellon University, announces the creation of the U.S.-Computer Emergency Response Team (US-CERT), a unit that is expected to grow by

2003
cont.

including other private sector security vendors and domestic and international CERT organizations.

Groups such as the National High-Tech Crime Unit (NHTCU) in the United Kingdom begin working with antivirus companies to find patterns in the coding of some of the most destructive Internet worms and viruses to determine whether they are the work of organized underground groups or other crime affiliates. NHTCU thinks that hidden somewhere in the lines of code are hints regarding the creator's identity and motives, and, possibly, imminent cyber-sabotage exploits.

In October, an international consortium releases a list of the top twenty Internet security vulnerabilities. The consortium—which includes the U.S. Department of Homeland Security, the UK National Infrastructure Security Coordination Center (NISCC), Canada's Office of Critical Infrastructure Protection and Emergency Preparedness (OCIPEP), and the SANS (SysAdmin, Audit, Network, Security) Institute—has as its objective the defining of an absolute minimum standard of security for networked computers.

In October, a French court finds the Internet search giant Google guilty of infringing Intellectual Property Rights. The company is fined 75,000 euros for allowing marketers to link their advertisements on the Net to trademarked search terms—a ruling that is said to be the first of this nature. The court gives the search company just one month to stop the practice.

On November 5, the news media report that a cracker has broken into one of the computers on which the sources of the Linux operating systems are stored and from which they are distributed worldwide.

On November 6, Microsoft Corporation takes the unusual step of creating a $5 million fund to track down malicious crackers targeting the Windows operating systems. That fund includes a $500,000 reward for information resulting in the arrest of the crackers who designed and unleashed Blaster and SoBigF. This Wild West–like bounty underscores the perceived problem posed by viruses and worms in a networked environment, as well as the difficulties associated with finding the developers. However, some cynical security critics say that the reward has more to do with Microsoft Corporation's public relations than it has to do with cyber-crime and punishment.

In November, the U.S. Federal Trade Commission (FTC) creates a national spam database and encourages people to forward to them all of the e-mail spam they receive. The FTC notes that in 2002, informants had reported more than 17 million complaints about spam messages to federal agents for investigation, and the FTC says that they receive nearly 110,000 complaints daily.

To control spam, on November 25, the U.S. Senate passes the CAN-SPAM Act of 2003, formally known as the Controlling the Assault of Non-Solicited Pornography and Marketing Act of 2003. Its purpose is to regulate interstate commerce in the United States by placing limitations and penalties on the transmission of spam through the Internet. Penalties include fines as high as $1 million and imprisonment for not more than five years for those found guilty of infringing the act. The act is to take effect on January 1, 2004.

2004   On January 21, the Recording Industry Association of America (RIAA) says that it has identified 532 song-swappers by the trails that their computers leave when the swappers illegally download music from the Internet. The swappers, identified by their IP addresses only, are targeted in four lawsuits, three filed in New York and one in Washington, D.C. The lawsuits are filed using the so-called John Doe process, which allows the RIAA to sue defendants whose names are not yet known.

On January 26, the worm W32.Novarg.A@mm, also known as MyDoom, spreads throughout the Internet and wreaks havoc. It arrives as an attachment with the file extension .bat, .cmd, .exe, .pof., scr., or .zip and

2004
cont.
affects Windows 2000, Windows 95, Windows 98, Windows Server 2003, and Windows XP systems but not the DOS, Linux, MacIntosh, OS/2, UNIX, or Windows 3.x systems. The damage done by MyDoom is estimated to be $2 billion worldwide.

In May, Sven Jaschan, aged eighteen, is arrested in Waffensen, Germany, in connection with creating and releasing on the Internet the Sasser worm. He later admits to police that he was the creator of Sasser.

On August 14, appearing before a judge in Seattle, Washington, eighteen-year-old Jeffrey Lee Parson concedes through a plea agreement that he had created the B variant of the Blaster worm, and that he used it to take over computers that were employed for an attack on nearly 50,000 other machines.

In mid-August, some would-be high-tech billionaire White Hats by the names of Sergey Brin and Larry Page commence their long-anticipated initial public offering (IPO) of Google, home of the world's leading Internet search engine.

From October to December, the Chinese authorities close more than 12,500 Internet cafes because they say that the cafes harm public morality by giving children access to violent games and sexually explicit material.

2005
In January, the U.S. FBI abandons its custom-built and highly controversial Internet surveillance technology known as Carnivore, designed to read e-mails and on-line communications of suspected cyber-criminals and terrorists. The FBI moves to commercially available software and encourages Internet providers to conduct wire-taps on suspected clients and then forward the information to the FBI.

On January 21, the confidential drug purchase histories of Harvard students and staff and the e-mail addresses of undergraduates who had been guaranteed nondisclosure by the university appear on Harvard's iCommons

Pool Tool. In fact, this information was accessible for months to anyone who had, say, a free Hotmail account and a few minutes' time to look up students' and staffers' eight-digit ID numbers. The vulnerability underscores the difficulty of securing the system when there is prevalent use of ID numbers to verify identity.

In February, while delivering a speech to security experts at the RSA Conference in San Jose, California, Microsoft Corporation founder Bill Gates says that the company will give away software to combat spyware, adware, and other privacy-invasive cyber nuisances.

In February, the 2004 National Technology Readiness Survey (NTRS) results indicate that spam, spread through the Internet, costs nearly $21.6 billion annually in lost productivity in the United States alone.

In February, a report released by a legislature committee finds that information on the Websites of New York's Department of Motor Vehicles, the New York Department of Education, the New York Department of Correctional Services, the State Division of Military and Naval Affairs, and the New York Power Authority has been cracked and defaced seventy-two times from 1999 to early December 2004. Jeff Klein, who heads the oversight committee and the report, notes that because state and private companies are not keeping important personal and homeland security information safe, identity theft can occur.

On February 15, in a plea agreement with prosecutors, Nicolas Jacobsen, aged twenty-two, pleads guilty in U.S. federal court in Los Angeles to one felony charge of intentionally gaining access to a protected computer and causing damage to it. His crime spree began in late 2003 and ended when he was arrested in the fall of 2004. Although Jacobsen's cyber-targets include Paris Hilton's T-Mobile Sidekick II, he is not known to be connected to the late February 2005 crack that resulted in Hilton's topless photos being shown on the Internet. Although most reports speculate that the leak was caused by an attack

2005
cont.
on T-Mobile's database (where the address book is backed up), others speculate that someone has access either to her phone or to her password.

On February 28, Phil Hollows, vice president of Security Products for OpenService, warns that from a Sarbanes-Oxley Section 404 perspective, any breach in Information Technology security poses a risk to a company's internal information system. Since IT underlies the critical business of recording and reporting all financial activity, a lack of control over IT security implies a lack of control by a company over its financial reports—a direct violation of section 404 of Sarbanes-Oxley. As the financial scandals at Worldcom and Enron have illustrated, having an auditor sign off is no guarantee that lawsuits can be avoided, and the Sarbanes-Oxley Act makes it very clear that CEOs and CFOs are personally liable for any material mispresentation.

On March 2, Harvard Business School says that it will reject 119 applicants who followed a cracker's instructions for peeking into the school's admission Internet Website to see if they had been accepted into the school—a month before results are to be disseminated. An anonymous man known as "brookbond," who says that he specializes in IT security, posted the instructions on Business Week Online's technology forum at 12:15 A.M. Harvard's admissions Website was, reportedly, vulnerable for more than nine hours.

On March 5, a cyber-war breaks out between Indonesia and Malaysia, brought on by a dispute over the Ambalat oil fields in the Sulawesi Sea and perhaps the termination of amnesty for illegal Indonesian workers.

On March 7, documents seized from three members of the Lashkar-e-Toiba (LeT) terrorist group killed in an encounter with Indian police indicate that they had planned to execute a "suicide" crack attack on the networks of companies having software and chip design operations in Bangalore and Karnal Singh. These companies include IBM, Intel, Texas Instruments, and Accenture.

On March 8, data security firms say that a new virus capable of attacking cell phones has emerged. F-Secure Corporation, a Finnish software security firm, says that the Commwarrior virus is the first one capable of spreading via multimedia messaging services containing photos, sound, or video clips.

On March 10, Trend Micro issues an alert regarding two new worms that spread through MSN Messenger, a widely used instant messaging platform. Known as kelvir.b and fatso.a, the worms have been reported on computers in the Asia Pacific region and in the United States.

On or about March 10, cyber-criminals stole passwords from legitimate users on as many as 32,000 Americans in an online database owned by the LexisNexis Group. Similar breaches occur at ChoicePoint, Inc., and at Bank of America. These cyber-crimes prompt calls for U.S. federal oversight through the General Services Administration (GSA) of the seemingly poorly regulated information brokerage industry.

On or about March 10, Limp Bizkit singer Fred Durst sues ten Website operators who posted his self-created sex tape on the Internet after it was stolen from his computer.

On March 10, according to a study commissioned by RSA Security, the huge growth in the use of wireless networks by businesses around the globe is making them increasingly subject to drive-by cracking. For example, in Europe's financial districts—the wireless networks seem to be growing at an annual rate of 66 percent. What is worrisome is that over a third of the businesses employing this technology are unprotected. By comparison, about 38 percent of the businesses in New York seem to be unprepared for such exploits, and about 35 percent of the businesses in San Francisco seem to be unprepared.

On March 16, five crackers in The Netherlands are found guilty of disabling a number of Dutch government

2005
cont.
Websites in 2004 because of DoS attacks. Citing protests against recent cabinet proposals as the motive for their crack attacks, the hacktivist group goes by the name of Oxlfe Crew. The spokesperson for the group, an eighteen-year-old who says that he would appeal his thirty-eight-day detention sentence, argues that there is no evidence to prove that he was involved in the attacks.

On March 21, Symantec Corporation issues a report saying that Internet attacks grew by 28 percent in the second half of 2004, compared with the first six months of 2004. On average, businesses and other agencies received 13.6 attacks on their computer networks daily in the second half of 2004—relative to 10.6 attacks in the first half of the year. Moreover, crackers seem to be setting their sights on mobile computers. The favored attack tools include phishing, spyware, and adware.

On March 21, survey findings released by security company Mazu Networks and the Enterprise Strategy Group show that almost half of 229 companies surveyed had a worm outbreak in 2004, despite their increased spending for network security features. Although approximately 75 percent of the IT professional respondents say that their companies increased security spending in 2004 to comply with Sarbanes-Oxley Act requirements, only 14 percent of the respondents report that they are confident that their networks could stop exploits.

On April 8, in a landmark case, spammer Jeremy Jaynes of North Carolina, described by prosecutors as being among the top spammers in the world, is sentenced to nine years in U.S. prison. This is the first successful felony prosecution in the United States for transmitting spam over the Internet.

On April 20, the Cyber Security Enhancement Act of 2005, or HR 285, is passed in the United States, specifying that the assistant secretary for Cyber Security will lead the directorate's National Cyber Security Division.

During the week of April 25, Microsoft Corporation's chair Bill Gates says that his company plans to put hardware computer security into a silicon chip instead of relying on only software in the next release of the Windows PC operating system, available in 2006.

During the first week of May, Websense's annual Web@Work survey indicates that 52 percent of the employee respondents admit that they would rather forgo their morning coffee than lose their ability to surf non-work-related Websites during the workday.

On May 23, Apple Computer, Inc., announces that they are in talks with the Intel Corporation about using the Intel chip in their Macintosh computer line—a prospect that could stir up the software and computer marketing world. Using Intel chips in the Macintosh computer makes it theoretically feasible for computer users to install Windows software on their "Mac" computers. With this announcement, Apple shares rise $1.94 to $39.49 on the Nasdaq Stock Market.

In early August, the Chinese search engine Baidu.com stock soars over 350 percent on the first day of trading. The thirty-seven-year-old founder, Robin Li, tells reporters that Chinese Internet use is still in its infancy, with just 94 million Internet users. Li has hopes of becoming another high-tech billionaire.

On September 8, a Massachusetts juvenile pleads guilty to cracking Paris Hilton's cell phone and dumping her personal information on the Internet. He also admits to cracking other systems in 2004, stealing personal information without authorization, and sending bomb threats to high schools in Florida and Massachusetts. All incidents occurred within a fifteen-month period, beginning March 2004. They allegedly have resulted in $1 million in damages to targets. The youth is sentenced to eleven months in a detention center.

On October 5, Chinese authorities shut down another two Internet Websites popular among academics, jour-

2005
cont.
nalists, and civil rights activists. One Website serves as an online discussion group to report on anticorruption protests in a southern village around Beijing. The other services ethnic Mongolians.

On November 17, the stock market shares of online search giant Google, Inc., cross the amazing $400-per-share mark.

2006
In January, more than twenty officers from Indian law enforcement agencies are trained in New Delhi in a Cyber Incident Response Course sponsored by the U.S. State Department's Anti-Terrorism Assistance Program.

In January, Chinese students and intellectuals express outrage at Beijing's decision to prevent access to Wikipedia, the fast-growing online encyclopedia offering more than 2.2 million articles in 100 languages. The online encyclopedia has emerged as an important source of scholarly information in China as well as many other developed and developing countries. The Website's independence on political issues involving Tibet and Taiwan has infuriated Chinese authorities.

In early February, the fascinating five-year Intellectual Property Rights legal battle between a U.S. company, NTP, and a Canadian company, RIM, is quickly coming to an end. The two companies are fighting about the patent rights to a wireless handheld called "The Black-Berry," which lets users read and respond to e-mail even when they are away from their computers and their desks.

In March, RIM, the Waterloo, Ontario, maker of the popular BlackBerry handheld device and enterprise server software, says it will pay its rival NTP U.S. $612.5 million to settle the five-year patent dispute.

In March, a Canadian team of musical buffs, an international aid group, and the state government of Chihuahua, Mexico, set up an Internet radio station, Radio-Tarahumara.com, aimed at bringing modern technology and communication to a remote Indian tribe in the mountains of Mexico to help them preserve their heritage. The team decided that helping the local Indian tribe compile educational and cultural digital content is a crucial step to introducing technology to the community and getting those who have not been exposed to technology interested in it. The Tarahumara Indians live in small adobe houses or caves, and there are only twelve computer-equipped sites in the entire region that can tune into the Internet radio station.

On March 14, a U.S. federal judge says that he intends to require Google, Inc., to turn over information to the Justice Department in its desire to revive a law making it more difficult for children to view online pornography. A legal showdown over how much of the World Wide Web's vast databases should be shared with the U.S. government has added fuel to the fires burning in the battle against the George W. Bush administration and Google, Inc. The company has resisted a subpoena to turn over any information because of user privacy and trade secret concerns. Although it defies the Bush administration, Google, Inc., has recently agreed to the Chinese government's censoring of its search results in China so that it can have better access to China's rapidly growing Internet market.

On March 22, the Microsoft Corporation offers free, unlimited technical support to competitors interested in making their software work with Microsoft servers. The company says that it wants to comply with a landmark European Union (EU) antitrust ruling. The EU placed a record $669 million fine against Microsoft Corporation in March 2004.

On September 5, in Odense, Denmark, police raided homes and detained nine men for allegedly preparing explosives for a terrorist attack in Denmark. The attacks were said to be related to Muslim outrage earlier in the year over media cartoons of the Prophet Mohammed.

5

Biographical Sketches

Tim Bray (b. 1955)

Tim Bray, a Canadian who attended the University of Guelph in Ontario, is credited with making online auctions like eBay possible. An entrepreneur, Bray worked for computer companies like DEC and then accepted the unique job in 1987 of managing the online transfer of the Oxford English Dictionary at the University of Waterloo in Ontario. He also founded the Open Text Company (a content management powerhouse) two years later, and by 1994, he had built the first successful Internet search engine.

Between 1996 and 1999, Bray was an independent consultant who was invited as an expert in the computer science field to the World Wide Web Consortium. This undertaking led to his eventually creating XML, or the Extensible Markup Language, which permits programmers to attach "tags" or universal codes to differentiate, say, a business name from a telephone number. The implication for e-commerce was great, because with this XML invention, purchase orders or invoices could be universally read as well as routed to the right applications. Put into context, as a key language of Web servers, the Extensible Markup Language facilitates the activities of an estimated 1 billion Internet users who are busy gaming online, buying things from Amazon, or partaking in eBay.

Although Bray wanted to be a high school math teacher after graduating from the University of Guelph, he said that there were no jobs in his then-chosen field, so he went into computer science. The rest of his career was all uphill from there, he admits. Bray now works for Sun Microsystems in Vancouver, British Columbia (Brearton, 2005; O'Reilly Media, Inc., 2006; Textuality Services, Inc., 2005).

Sergey Brin (b. 1973) and Larry Page (b. 1973)

Sergey Brin, born in 1973 in Russia, and Larry Page, born the same year in the United States, are two Stanford University Ph.D. dropouts who started Google, Inc., the six-year-old Internet search engine that is now valued at $120 billion.

Before Brin and Page launched Google, Inc., Internet searches often returned more useless than useful information. Users would have to wade through all of the useless information to find something worthwhile—a huge waste of resources and time. Internet pioneers Brin and Page streamlined the search process to such a degree that today, the Google search engine receives, on average, 200 million queries daily (A & E Television Networks, 2006; Farzad and Elgin, 2005).

Brin's father was a Russian professor who moved to teach at the University of Maryland, where he is still employed. Brin's mother works for NASA. When Brin was nine years old, his father apparently gave him his first computer, a Commodore 64. Both Brin and Page went to Montessori schools, where, they say, their creative and free-thinking spirits were fueled.

Page says that when he and Brin were working on their Ph.D.s, they had no interest in starting a business, but when their computer science research turned up the promise of a faster kind of search engine, they decided to abandon academic pursuits (and a likely smaller pay check) for the entrepreneurial spirit. Page's father, by the way, was also a university professor of computer science, but at Michigan State University (Woopidoo.com, 2006).

According to *Forbes* business magazine, Brin and Page are each now worth about $11 billion and are ranked the second richest Americans under the age of forty—after the founder of eBay, Pierre Omidyar (Blinkbits, 2006).

Like other billionaires who have made their success through e-commerce, Brin and Page are interested in giving back to the collective good of society. To that end they recently announced that Dr. Larry Brilliant, a former high-tech executive and physician who specializes in global health, will head the Google.org philanthropic arm. The sixty-one-year-old Dr. Brilliant is currently director of the Seva Foundation in Berkeley, California, which he founded in 1979 to fight blindness in people in the developing nations. In his new position with Google, Dr. Brilliant will have to decide how to invest about 1 percent of the company's annual profit and 1 percent of Google's annual stock—which amounts to about $ 1 billion—for the betterment of society. Dr. Brilliant said that he will likely promote public health (Delaney, 2006).

Tim Burners-Lee (a.k.a. Sir Timothy "Tim" John Burners-Lee) (b. 1955)

A graduate of Oxford University in England, Tim Burners-Lee in 1989 invented the World Wide Web, an Internet-based hypermedia start-up for information sharing on a wide-ranging scale. This brilliant scholar currently holds the 3Com Founder's Chair at the Laboratory for Computer Science and Artificial Intelligence (AI) at MIT. He also directs the World Wide Web Consortium, a group of companies and agencies interested in capitalizing on the World Wide Web's potential. A year after he invented the World Wide Web, he wrote the first Web client (browser-editor) and server. In 1999 the book by Tim Burners-Lee's (along with coauthor Mark Fischetti), entitled *Weaving the Web*, was released by Harper Publishers, San Francisco (www.w3.org, 2005).

Vinton Cerf (a.k.a. Father of the Internet) (b. 1943)

Known to many as "the Father of the Internet," Vinton Cerf was the codesigner, along with Robert Kahn, of the TCP/IP protocols and the architecture of the Internet. For that major accomplishment, in 1997 the pair were presented by President Bill Clinton with the U.S. National Medal of Technology. In 2004 the pair were also honored with the ACM Alan M. Turing Award, considered by many in the computer science field to be the equivalent of the Nobel Prize.

Today Vinton Cerf is employed in senior management for Google, Inc., the online search engine. He identifies new enabling technologies to support the development of the next evolution of products and services for Google. He is also chair of the Board of the Internet Corporation for Assigned Names and Numbers (ICANN).

Earlier in his career, from 1976 to 1982, Cerf played a critical role in leading the development of the Internet, as well as in creating data packet and security technologies when he was employed by the U.S. Department of Defense's Advanced Research Projects Agency (ICANN, 2005).

Stephen Cook (b. 1939)

Born in the United States in 1939, Stephen Cook, a distinguished university professor now at the University of Toronto in Canada, received his graduate degrees from Harvard University. A mathematics professor, Dr. Cook published what some believe was quite an esoteric work known as *The Complexity of Theorem Proving Procedures* in 1971. Its shorter title was *Cook's Theorem.* The wonderful thing about this theorem is that it identified a large group of computational search problems that would normally take even the most powerful computers millions of years to compute. This theorem has been an especially strong contributor to the important field of cryptography. Without cryptography, online transactions as we now know them would not be secure.

Without getting into all of the technical details, the theorem's main contribution was that the security of every e-commerce transaction depends on a rather unique silicon irony—namely, that the relative safety of users' credit card information, say, relies not so much on what computers *can do,* but rather on what they *cannot do.* Dr. Cook is the only professor in Canada to have won the Association for Computing Machinery's A.M. Turing Award, considered by many to be computer science's most prestigious prize (Brearton, 2005; NSERC, 2006).

Shawn Fanning (b. 1980)

At age eighteen, Shawn Fanning was a creative, entrepreneurial young adult who was the lead software engineer and cofounder—along with Sean Parker—of Napster, a controversial but free software program that during its short life span let online users exchange MP3 music files without having to pay royalties to the artists and without having to purchase CDs at stores. Napster, in fact, could best be described as a "music bootlegging" business. Fanning says that the creative idea for Napster was conceived while he was a first-year university student at Northeastern University in Boston.

Not too long after Napster's release, the big music companies—including Universal Studios and Sony—alleged that Fanning's technology was an infringement of the U.S. Digital Millennium Copyright Act of 1998, and that he was cheating artists out of their royalties. As a result, Napster was eventually pushed out of business. They filed for bankruptcy in 2002 (Ante, 2000).

Two years later (on December 3, 2004), Shawn Fanning launched Snocap, a music-licensing platform that permitted music download services by Internet-connected users—but with the users' knowing that they were paying music copyright holders their share of the cost. In short, Snocap acted as a licensing manager. Upon the announcement of Snocap's release by Fanning, an online service in the United Kingdom called Wippit claimed that it used a similar system for its music-sharing service and that it was the real pioneer in this space, not Fanning (Smith, 2004).

It is interesting to note that Fanning is quite a legend among youth, particularly among young hackers. In February 2005, Paramount Pictures and MTV films partnered to hire Alex Winter to write "Napster: The Shawn Fanning Bio Project." Although in 2002 MTV first developed the media project about Fanning as a TV movie, the company later opted, instead, for a "big screen" movie version of this creative man's biography (MovieWeb, Inc., 2005).

David Filo (b. 1967) and Jerry Yang (b. 1969)

The Yahoo online search engine, now worth $60 billion, was started by two electrical engineering Ph.D. candidates at Stanford University: David Filo and Jerry Yang. Like Brin and Page, these two creative entrepreneurs had not seriously thought about making business their careers. Actually, the search engine started in a campus trailer in February 1994, with the main purpose of helping those two men keep track of their personal interests on the Internet. But because the self-selected Internet task was somewhat more interesting than their doctoral study tasks, the pair apparently started spending more time on their lists of favorite links than on their dissertations. When the pair's lists became too large and unwieldy, they put them into categories and subcategories—which is how the framework behind Yahoo was created. Although Filo and Yang originally called their Website "Jerry and David's Guide to the World Wide Web," they eventually changed the name to Yahoo—which means Yet Another Hierarchical Officious Oracle. Since the pair of creators had a sense of humor, they wanted their Website to have one, too—which is why they came up with the name. They also wanted the name to indicate an entity that was somewhat rude, unsophisticated, and somewhat uncouth—just like its developers, they jested (Farzad and Elgin, 2005; Yahoo, Inc., 2006).

William H. Gates (b. 1955)

Microsoft Corporation's chair and chief software architect, Bill Gates founded the company with childhood friend Steve Allen back in 1975. For the year ending 2005, the Microsoft company had business revenues of nearly $40 billion.

An "eager beaver" in terms of computing, Gates apparently started programming at age eleven. He started Harvard University in 1973—where he met Steve Ballmer, who now has the prestigious position of chief executive officer (CEO) of Microsoft. In order to run the Microsoft business more effectively, Gates chose to drop out of Harvard University in his third year.

Forever devoted to computing, Mr. Gates in 1999 wrote a book entitled *Business @ the Speed of Thought*. The focus of the book was how computer technology can solve complex business problems in unique ways. The book itself became a best-seller. It made it to the *New York Times* best-seller list—and held the top position for a solid seven weeks. The book was sold in more than sixty countries and was translated into at least twenty-five different languages.

A philanthropist as well as an entrepreneur and visionary, Bill Gates today gives away the proceeds of his books to non-profit organizations promoting the use of technology in education. The company itself is committed to promoting programs advancing the use of technology for children and young adults (Microsoft Corporation, 2006)

James Gosling (a.k.a. Software Wizard, a.k.a. World's Greatest Programmer) (b. 1955)

Born in Calgary, Alberta, Canada, James Gosling headed a talented group at Sun Microsystems that developed a programming language and environment known as Java. Java's primary asset was that it could convert the one-dimensional and very noninteractive Web into a system permitting software on any operating system (that is, Microsoft Windows, UNIX, or Macintosh) to communicate by streaming bits of information that was universally translatable. The highly interactive World Wide Web that adults and children experience today is as user-friendly as it is because of this wonderful invention called Java. It is important to note that the reputation of Sun Microsystems also rose when the company permitted other software developers to make use of the Java programming language free of charge. A white paper on Java, written by Gosling, can be found at: <http://inventors.about.com/gi/dynamic/offsite.htm?site=http://java.sun.com/docs/white/langenv/>.

When recently asked if Java is at the mature end of its evolutionary phase, Gosling remarked: "It feels like we're only a third of the way through exploiting into what we can really do with Java. There's a lot of play in this puppy yet" (LaMonica, 2006).

James Gosling received his undergraduate degree from the University of Calgary in Canada in 1977 and his Ph.D. in 1983 from Carnegie Mellon University in the United States. He began employment with Sun Microsystems in 1984 and is still employed there, but now with the important title of chief technical officer (Brearton, 2005).

Steve Jobs (b. 1955)

Steve Jobs, one of the two cofounders of Apple Computer, Inc., seems to have had a flexible, broad-based intelligence even in his early adulthood. He studied physics, literature, and poetry while at Reed College in Portland, Oregon. In 1976, Jobs's entrepreneurial drive became evident, as he sold his Volkswagen minibus and used the proceeds to start his computer company, along with fellow cofounder Steve Wozniak.

In 1980 the pair took their company public at $22 per share, and within four years they had created the MacIntosh computer. During the years 1986 through 1997, Jobs founded NeXT Software, Inc., a company that made hardware to exploit object-oriented technologies (that is, those that model features of the real world using computer programming). In 1997, Jobs sold the company to Apple, where he is currently the CEO.

But Steve Jobs's entrepreneurial activities did not stop there. In 1986 he cofounded an animation company called Pixar Animation Studios—the producer of many high-grossing films such as *Toy Story, Toy Story 2, A Bug's Life, Monsters, Inc., Finding Nemo,* and *The Incredibles.* To date, Pixar's six films have brought in more than $3 billion in revenues at the worldwide box office. The CEO of this Academy Award–winning, creative company is Steve Jobs.

Since 1997, Steve Jobs has helped Apple Computer, Inc., bring to market creative and high-demand products, such as the iPod portable music player and the itunes online music outlet (Apple Computer, Inc., 2004; Schell and Martin, 2004).

Johan Johansen
(a.k.a. DVD-Jon) (b. 1983)

Johan Johansen was born to a Polish mother and a Norwegian father in 1983—the year, he maintains, that he had his first encounter with a computer (his father's Sinclair ZX Spectrum). Obviously, Johansen was a bright child, for in January 2000 he received the Karoline Award in high school for his excellent grades (he had a 5.75 average of a 6.0 maximum) and his contributions to society—in art, culture, and the like. However, bored with a structured high school regimen and called by a creative spirit, Johan dropped out of high school in June of the same year.

A year earlier, in 1999, he, along with two European computer programmers, had coauthored the DeCSS decryption program. DVD-Jon, as he eventually became known, was only fifteen years old at the time. The moniker derived from an exploit that occurred in 1999–2000. DVD-Jon cracked a DVD-access code and published his decryption program on the Internet. For that action he was sued by the U.S. DVD Copy Control Association (DVD-CCA) and the Norwegian Motion Picture Association of America (MAA). Although he was charged by the Norwegian police with computer crime for the act, two courts in Norway unanimously ruled that DVD-Jon's decryption program did not breach Norwegian law. Therefore, he was eventually acquitted of all charges.

In 2002, Johansen accepted the Electronic Frontier Foundation Pioneer Award for his contribution toward the development of DeCSS. Johan Johansen continues to produce creative software, and in October 2005 he moved to San Diego, California, so that he could be closer to other creative types with similar interests (Johansen, 2004; Vaagan and Koehler, 2005).

Mers Kutt (b. 1933)

Mers Kutt is the Canadian inventor of the first Personal Computer (PC), the MCM/70. Believe it or not, this MCM/70 has now passed its thirtieth anniversary of user demand. Born in Winnipeg, Manitoba, Mers is a mathematician and an entrepreneur who in 1973—two years before Bill Gates founded Microsoft—founded Toronto-based Micro Computer Machines, Inc.

The MCM/70 was a small desktop microcomputer designed to provide the APL programming language environment for various applications—scientific, educational, and business. The company that Kutt founded was one of the first to recognize the power of, and to capitalize on, the microprocessor technology's huge capabilities for developing a whole new generation of cost-effective computing machines.

The MCM/70 was apparently a vision planted in Kutt's mind back in the late 1960s when he was a mathematics professor at Queen's University in Kingston, Ontario. He thought that what the world needed was not another big, clunky computer or another fan-driven minicomputer but a small, powerful machine that could be for individuals' personal use. What was needed, in essence, was a Personal Computer, or PC. Not only was Kutt's MCM/70 quieter and more powerful than the loud, fan-driven minicomputers in use at the time, but it had some advanced features built into it that even IBM's PC was lacking.

Kutt's other inventions included the key edit, a data preparation system for computers that evolved after the IBM punch card, prevalently used in the late 1960s. What this device did was to permit a computer user to enter data into a computer. The computer would then edit the information that had been inputted (Canadian Information Processing Society, 2003; Stachniak, 2003).

Mafiaboy (b. 1985)

Mafiaboy, likely Canada's most famous cracker, was a minor (fifteen years of age) when in February 2000 he cracked Internet servers and used them to launch very costly Denial of Service (DoS) attacks on several high-profile e-commerce Websites, such as Amazon, eBay, and Yahoo. After pleading guilty on January 18, 2001, to the charges, he was sentenced by a Montreal judge to eight months in a youth detention center, and he was fined a mere $250.

After his release, Mafiaboy appears to have reflected and acted on some lifestyle changes. Today he uses his technical and creative talents as a columnist for CNews Tech News in Montreal. Still going by the moniker Mafiaboy, his recent columns have had such interesting high-tech titles as: "Know the Risks When Purchasing Online," "Want Free Distance? Use the Internet," "Hacking Becoming Even Easier," and "Can a Printer Invade Your Privacy?" (Canoe, Inc., 2006; Schell and Dodge, 2002).

Kevin Mitnick (a.k.a. Condor) (b. 1963)

Kevin Mitnick, a cracker who was hunted down by the Federal Bureau of Investigation (FBI) for cracking computers in his early adulthood, for stealing corporate secrets, for scrambling telephone networks, and for cracking the U.S. national defense warning system, seems now to have been rehabilitated.

Besides being a regular speaker at the Black Hat and DefCon security gatherings held in Las Vegas every July, Matnick now owns his own IT security company—Mitnick Security Consulting, LLC. The Website for his company says that he offers a comprehensive range of IT security services to help companies protect their valuable assets, and he has lots of contacts in the Information Technology Security business.

Mitnick is quite the icon in the Computer Underground. It is interesting to note that in February 1995, White Hat hacker Tsutomu Shimomura helped the FBI locate Mitnick for his series of crack exploits. As a result, Mitnick was imprisoned on a twenty-five-count indictment that included charges related to wire fraud as well as illegal possession of computer files stolen from well-known high-tech companies such as Sun Microsystems and Nokia. Mitnick was released from prison in 2000.

Mitnick has now turned his creative energy to writing security books, one of which is *The Art of Deception: Controlling the Human Element of Security*—a book that details the importance of social engineering in successful crack attacks. His latest book release is *The Art of Intrusion: The Real Stories behind the Exploits of Hackers, Intruders, and Deceivers* (Mitnick Security Consulting, LLC, 2004).

Pierre Omidyar (b. 1967)

On Labor Day in 1995, Pierre Omidyar decided to launch eBay as an experiment—one that has since paid off extremely well. Today the online auction eBay is worth about $65 billion (Farzad and Elgin, 2005).

Pierre graduated from Tufts University in 1988 with an undergraduate degree in computer science (Omidyar Network, 2004). At the time that he was contemplating the experiment, Pierre asked himself, "What would happen within a marketplace if everyone had equal access to information and tools?" He quickly found his answer. There are voluminous benefits for online users when there is an open and secure environment in which to conduct business transactions. Today, hundreds of thousands of individuals lead economically sustainable lives because of eBay, an online auction. Besides participating in an online auction, they are business people in their own right. According to eBay's founder, more than 150 million people trust strangers with each transaction.

It is interesting to note that the billionaires who have made their fortunes on the Information Highway seem to have developed a real commitment to giving back to the community for the collective good of society. It is, perhaps, an extension of the White Hat hacker ethic applied to business.

For example, after eBay's success and its IPO in 1998, Omidyar and his wife, Pam, cofounded the Omidyar Foundation to fund nonprofit organizations. In short, eBay's tremendous social impact as a for-profit company has demonstrated that businesses around the globe could also be effective tools for making the world a better place to live in and for redistributing the wealth.

In June 2004, the Omidyars created a new entity called the Omidyar Network to invest in for-profit, nonprofit, and public policy efforts. So far, the Network (even its name has Internet implications) has funded a number of social areas so that—according to the network's Website—"more and more people discover their own power to make good things happen." These social areas have included microfinance, citizens' journalism, open-source software development, and IP protection. The Omidyars say that over the next few years, the network will invest $400 million across sectors and will search for economically sound financial models that will only evolve even further—if they result in positive social impacts (Omidyar Network, 2004).

Dennis Ritchie (a.k.a. dmr) (b. 1941)

Dennis Ritchie, who is still at the Computing Sciences Research Center at Bell Laboratories but under the Lucent Technologies label, joined the company in 1967—back when the company was still known as the American Telephone and Telegraph Company (or AT&T). Before long Dennis Ritchie met Ken Thompson at the Research Center, and the pair, along with other creative talents at the lab, created UNIX—an open operating system for minicomputers. Besides helping UNIX users with general computing, word processing, and networking, UNIX also soon became a standard computer language.

In 1971, Ritchie wrote the *UNIX Programmer's Manual, First Edition,* and in 1990 he helped write the UNIX *Tenth Edition Manual,* the last UNIX manual published by his group.

Also the creator of the popular C programming language, Dennis Ritchie received his graduate degrees in applied mathematics from Harvard University. Dennis Ritchie is currently a researcher at Lucent Technologies in Murray Hill, New Jersey (Bell Labs, 2002; Ritchie, 2006).

Dmitry Sklyarov (b. 1974)

In July 2001 the Russian-born computer programmer Dmitry Sklyarov made headlines when he was arrested at the DefCon 9 hacker convention in Las Vegas. He had been about to speak on a software package that he had developed for Elcomsoft, his Russian employer.

What bothered FBI agents as they handcuffed Sklyarov and whisked him off to jail was that Sklyarov's software permitted Internet users to convert the so-called copy-protected Adobe eBook file format to a more commonly utilized and freely copyable computer file. Said to have infringed the U.S. Digital Millennium Copyright Act (DMCA) of 1998, Sklyarov was imprisoned by the federal agents who had cut his appearance at the hacker convention short.

Soon after his arrest, the Electronic Frontier Foundation (EFF) got involved in Sklyarov's case and lobbied on his behalf. The EFF claimed that jurisdictional issues applied and that Sklyarov's activities were perfectly legal in his homeland—Russia. As the case went on, Sklyarov's employer, Elcomsoft Co., Ltd., was also charged with infringing the DMCA. In December 2002 both the computer programmer and his company were cleared of any wrongdoing by the courts, and on December 13, 2001, the creative doctoral student returned home to be with his wife and two children (Glasner, 2002).

Ken Thompson (b. 1943)

Dennis Ritchie's coinventor of the UNIX operating system, Ken Thompson, is now retired from Bell Labs Computing Research Center (now under the Lucent Technologies umbrella). This creative spirit now lives in Murray Hill, New Jersey.

Thompson arrived at Bell Labs in 1966. Within three years, he worked with Dennis Ritchie to produce UNIX. In 1970,

Thompson also focused on creating the B programming language—which was the precursor to Dennis Ritchie's C language. Then, in 1973, Thompson rewrote UNIX in C.

In 1995 and 1996, Thompson took up a visiting professorship at the University of California, Berkeley, where he received his undergraduate and master's degrees in electrical engineering. In 1998, near the end of his career, Thompson was awarded, along with Dennis Ritchie, the National Medal of Technology for his key role in developing UNIX. Two years after receiving that honor, he opted to retire (Lucent Technologies, 2002).

Ray Tomlinson (b. 1941)

In 1971–1972, Ray Tomlinson helped contribute to one of the greatest communication gifts in Internet history: electronic mail, or e-mail. Tomlinson, one of the forefathers of the Internet, worked on ARPAnet. A Harvard University graduate, Tomlinson is the creative person who chose the @ sign for e-mail—a sign still in use today. The sign simply means "at" (Gaudin, 2002).

It is important to note that Tomlinson was not actually the inventor of e-mail, for it had been around since about 1965, when Fernando Corbato and his colleagues at MIT developed a software program allowing users of MIT's Compatible Timesharing System (CTSS) to share messages. The shortfall was that the program let individuals using only one machine communicate with one another.

Tomlinson's value-added contribution was that he made it possible for users to exchange messages between machines and in different locations—even if they were located in different continents or across oceans. E-mail—as users today know it and love it—was thirty years old on October 8, 2001 (BBC NewsOnline, 2001).

Steve Wozniak (a.k.a. Oak Toebark; a.k.a. The Wizard of Woz; a.k.a. The Other Steve) (b. 1950)

Steve Wozniak, a creator and entrepreneur of many names, has lived up to his Leo horoscope sign by being one of the world's best-known high-tech leaders. Together with Steve Jobs, Steve Wozniak made his fortune in the high-tech market with the invention of the Apple II computer more than twenty-five years ago.

An engineer by profession, Wozniak was a creative sort from the tender age of eleven, when he built his own ham radio station. Two years later he began designing computers, and when he went to university—where he met Steve Jobs—the pair had fun creating "blue boxes." These permitted Jobs and Wozniak to phreak (that is, manipulate) long-distance telephone lines to get telephone calls for free. In fact, at the 2004 Hackers on Planet Earth gathering in New York City, the Wizard of Woz, as he is called, joked with the audience about those fond "blue box" exploit memories from the good old days.

Entrepreneurial Steve Wozniak and Steve Jobs—the two Steves—in 1976 started the Cupertino, California, Apple Company, and four years later, the company went public—creating two very wealthy men. As a result of a power struggle between the two entrepreneurs, both men left the Apple Company in 1985. Although Steve Jobs came back to Apple as CEO in 1997—and has continued on in that capacity—Steve Wozniak has stayed away from the company since that time, although he is still on Apple's payroll.

With the entrepreneurial spirit still within him, Steve Wozniak went on to found the Electronic Frontier Foundation (EFF) advocacy group, the Tech Museum, and the Children's Discovery Museum of San Jose. In 2000 this philanthropic individual was awarded the highly regarded Heinz Award for Technology for designing the world's first PC and the Economy and Employment Award for later sharing his passions for mathematics and electronics with elementary school students. During 2000 he was also inducted into the Inventors Hall of Fame (Nowak, 2006; Unuson Corporation, 2003).

References

A & E Television Networks. "Rich List: Google Boys (Sergey Brin and Larry Page)." The Biography Channel, http://www.thebiographychannel.co.uk/biography_story/1352:1491/1/Google_Boys_-_Sergy_Brin_Larry_Page.htm (accessed February 23, 2006).

Ante, Spencer. "Shawn Fanning." Business Week, http://www.businessweek.com/2000/00_20/b3681054.htm (accessed May 15, 2000).

Apple Computer, Inc. "Steve Jobs." Apple Computer, http://www.apple.com/pr/bios/jobs.html (accessed December, 2004).

BBC NewsOnline. "Indepth: Dot Life; H@ppy Birthday to You." BBC News, http://news.bbc.co.uk/1/low/in_depth/sci_tech/2000/dot_life/1586229.stm (accessed October 8, 2001).

Bell Labs. "Dennis M. Ritchie." Bell Labs, http://www.cs.bell-labs.com/who/dmr/ (accessed March, 2002).

Blinkbits. "Sergey Brin Wikipedia RSS Feed." Blinkbits, http://www.blinkbits.com/bits/viewtopic/sergeibrin_sergey_brin_wikipedia_rss_feed?t=5785642 (accessed February 14, 2006).

Brearton, Steve. "Innovation." *Globe and Mail Report on Business*, September 22, 2005, pp. 54–55, 57, 59, 63, 65–66, 68.

Canadian Information Processing Society (CIPS). "CIPS Honours Mers Kutt, Inventor of First PC." CIPS, http://www.cips.ca/news/national/news.asp?aID=1747 (accessed October 31, 2003).

Canoe, Inc. "Mafiaboy." Canoe, http://cnews.canoe.ca/CNEWS/Tech News/Mafiaboy/ (accessed February 23, 2006).

Delaney, Kevin. "Dr. Brilliant: From Rock-Star Physician to Advisor for Google." *Globe and Mail*, February 22, 2006, p. B15.

Farzad, R., and B. Elgin "Googling for Gold." *Businessweek* 49 (December 5, 2005): 48–52, 54.

Gaudin, Sharon. "A Conversation with the Inventor of Email." IT Management, http://itmanagement.earthweb.com/entdev/article.php/1408411 (accessed July 16, 2002).

Glasner, Joanna. "Elcomsoft Found Not Guilty on All Counts!" Radar News, http://www.onlisareinsradar.com/archives/000747.php (accessed December 17, 2002).

ICANN (Internet Corporation for Assigned Names and Numbers)."Vinton G. Cerf." ICANN, http://www.icann.org/biog/cerf.htm (accessed September 27, 2005).

Johansen, Johan. "So Sue Me: About Me." Johansen, http://nanocrew.net/about/ (accessed 2004).

LaMonica, Martin. "Is Java Getting Better with Age?" CNET Networks, Inc., http://www.builderau.com.au/program/java/soa/Is_Java_getting_better_with_age_/0,39024620,39228342,00.htm (accessed January 11, 2006).

Lucent Technologies. "Kenneth Thompson." Lucent Technologies, http://www.bell-labs.com/about/history/unix/thompsonbio.html (accessed 2002).

Microsoft Corporation. "William H. Gates." Microsoft Corporation, http://www.microsoft.com/billgates/bio.asp (accessed February 23, 2006).

Mitnick Security Consulting, LLC. "Mitnick Security Consulting, LLC." Mitnick Security, http://www.kevinmitnick.com/index.php (accessed 2004).

MovieWeb, Inc. "MTV Hires Alex Winter to Pen Napster: The Shawn Fanning Bio." Movie Web, http://www.movieweb.com/news/65/6965.php (accessed February 25, 2005).

NSERC (Natural Sciences and Engineering Research Council). "Stephen Cook." NSERC, http://www.nserc.ca/news/2006/p060214_cook.htm (accessed February 23, 2006).

Nowak, Peter. "What the Other Steve Is Saying about Apple's Striking Resurgence." *Globe and Mail*, February 23, 2006, p. B10.

Omidyar Network. "Pierre Omidyar." Omidyar Network, http://www.omidyar.net/corp/t_pierre.html (accessed 2004).

O'Reilly Media, Inc. "Tim Bray." O'Reilly Media, Inc., http://www.xml.com/pub/au/10 (accessed February 23, 2006).

Ritchie, Dennis. "Dennis M. Ritchie." Dennis Ritchie Website, http://cm.bell-labs.com/cm/cs/who/dmr/bigbio1st.html (accessed February 23, 2006).

Schell, B., and J. Dodge, with S. Moutsatsos. 2002. *The Hacking of America: Who's Doing It, Why, and How.* Westport, CT: Quorum.

Schell, B., and C. Martin. 2004. *Cybercrime: A Reference Handbook.* Santa Barbara, CA: ABC-CLIO.

Smith, Tony. "Shawn Fanning's Snocap Touts Vision of P2P Heaven." The Register, http://www.theregister.co.uk/2004/12/03/snocap_launch/ (accessed December 3, 2004).

Stachniak, Zbigniew. "The Making of the MCM/70 Microcomputer." IEEE, Inc., http://csdl2.computer.org/persagen/DLAbsToc.jsp?resourcePath=/dl/mags/an/&toc=comp/mags/an/2003/02/a2toc.xml&DOI=10.1109/MAHC.2003.1203059 (accessed April–June 2003).

Textuality Services, Inc. "Tim Bray." Textuality Services, Inc., http://www.textuality.com/ (accessed 2005).

Unuson Corp. "Short Bio for Steve Wozniak." Unuson Corporation, http://www.woz.org/wozscape/wozbio.html (accessed May 14, 2003).

Vaagan, R., and W. Koehler. "Intellectual Property Rights vs. Public Access Rights: Ethical Aspects of the DeCSS Decryption Program." Information Research, http://informationr.net/ir/10–3/paper230.html (accessed April 2005).

Woopidoo.com. "Larry Page Biography." Woopidoo, http://www.woopidoo.com/biography/larry-page/ (accessed February 23, 2006).

www.w3.org. "Tim Berners-Lee." W3, http://www.w3.org/People/Berners-Lee/ (accessed December 31, 2005).

Yahoo, Inc. "The History of Yahoo! How It All Started." Yahoo, Inc., http://docs.yahoo.com/info/misc/history.html (accessed February 24, 2006).

6

Data and Documents

Chapter 6 discusses newsworthy Internet crime case facts prosecuted in the United States under the computer crime statute 18 U.S.C. Section 1030. In the United States, the Computer Fraud and Abuse Act (CFAA) has been the primary federal statute criminalizing Internet abuses regarding privacy, security, and trust; to strengthen its powers, it was modified in 1996 by the National Information Infrastructure Protection Act and codified at U.S.C. subsection 1030, Fraud and Related Activity in Connection with Computers.

Following the September 2001 terrorist attacks on the World Trade Center and the Pentagon, the U.S. government began to pass a number of laws not only to curb Internet crimes but also to cope with potential terrorist and cyber-terrorist activities in the United States. The first half of this chapter summarizes legislation pertaining to the Internet, as well as ten case facts prosecuted in the United States under the computer crime statute U.S.C. section 1030 for the past five years. The focus of this section is on specific and highly publicized case facts of harm to persons and to property: child pornography, identity theft, worms and viruses, Intellectual Property infringement, and online fraud.

The second half of this chapter presents interesting data regarding Internet prevalence and use, taken from a global perspective.

Internet Legislation in the U.S.

18 U.S.C.

As of this writing in 2006, the following statutes pertain to Internet crime in the United States under 18 U.S.C.:

191

Section 1020: Fraud and related activity regarding access devices.

Section 1030: Fraud and related activity regarding computers.

Section 1362: Communication lines, stations, or systems.

Section 2510: Wire and electronic communications interception as well as the interception of oral communication.

Section 2512: The manufacture, distribution, possession, and advertising of wire, oral, or electronic communication intercepting devices prohibited.

Section 2517: Authorization for disclosure and use of intercepted wire, oral, or electronic communications.

Section 2520: Recovery of civil damages authorized.

Section 2701: Unlawful access to stored communications.

Section 2702: Voluntary disclosure of customers' communications or records.

Section 2703: Required disclosure of customers' communications or records.

Section 3121: Recording of dialing, routing, addressing, and signaling information.

Section 3125: Emergency pen register and trap and trace device installation.

Health Insurance Portability and Accountability Act of 1996 (HIPAA)

HIPAA, short form for the Health Insurance Portability and Accountability Act of 1996, is focused on health protection for U.S. employees in a number of ways, with the Centers for Medicare and Medicaid Services (CMS) having the responsibility to implement various unrelated provisions of HIPAA.

Title I of HIPAA maintains that health insurance coverage for individuals and their families will carry on when they transfer or lose employment, and Title II requires the Department of Health and Human Services to develop and maintain national standards for e-transactions in health care. Title II also speaks to the security and privacy of online health data, capable of being sent through the Internet.

The developers of HIPAA felt that such standards would improve the efficiency and effectiveness of the U.S. health care system by encouraging the secure and private handling of electronic data. From an information security perspective, it is interesting to note that HIPAA requires a double-entry or double-check of data entered by personnel.

With a deadline of April 21, 2005, all U.S. healthcare organizations had to meet the new HIPAA Security Rule regulations by taking extra measures to secure protected health information. The final version of the Security Rule was published on April 21, 2003 (Centers for Medicaid and Medicare Services, 2002).

Digital Millennium Copyright Act of 1998 (DMCA)

The protection of Intellectual Property Rights (IPR) from attack by cyber-criminals is, for many modern-day businesses, as important as dealing with crack attacks on computer networks.

Enacted in October 1998, the DMCA was intended to implement under U.S. law certain worldwide copyright laws to cope with emerging digital technologies by providing protection against the disabling or the bypassing of technical measures designed to protect copyright. The DMCA sanctions apply to anyone who attempts to impair or disable an encryption device protecting a copyrighted work, typically using the Internet. The Website for the DMCA is http://www.copyright.gov/legislation/dmca.pdf.

Gramm-Leach Bliley Act of 1999 (Financial Services Modernization Act)

Personal information that many citizens would consider to be private—such as their bank account numbers and their bank account balances—is routinely exchanged for a price by banks and credit card companies. For that reason, the Gramm-Leach-Bliley Act (GLBA), or Financial Services Modernization Act of 1999, brought in some privacy protections against the sale of citizens' private information of a financial nature. Also, the GLBA codified protections against "pre-texting," defined as the act of getting someone's personal data through false means.

The purpose of the GLBA was to remove regulations preventing banks, insurance firms, and stock brokerage firms from merging. However, argued critics, if such regulations were removed, merged financial institutions would be able to have access to a huge quantity of citizens' personal information, typically stored on networks and capable of being sent through the Internet—with little or no restrictions on how the personal information could be used. Before the passage of the GLBA, an insurance company having citizens' health records, for example, would be distinct from, say, a banking institution that had personal information on clients wanting a house mortgage. With the passage of the GLBA and following the merger of two such firms, they could not only pool the information they had on all of their clients but also sell it to interested third parties.

Because of these risks, the GLBA included three requirements to protect the personal data of individuals: (1) information had to be securely stored; (2) the merged institutions had to advise clients about the policy of sharing personal financial information with others; and (3) the institutions had to give consumers the right to opt out of the information-sharing schemes if they so desired (Electronic Privacy Information Center, 2005).

It is generally not well known that a Victoria's Secret catalog—an Internet e-commerce Website and regular retail outlet for women's lingerie—is one of the key reasons that Congress included privacy protections for financial information when passing the Gramm-Leach-Bliley Act (GLBA). The debate started in Congress when Representative Joe Barton talked about his concerns that his credit union had sold his address to Victoria's Secret, after which, he said, he started getting the catalogs at his home in Washington. He noted that he became stressed when the catalogs starting arriving, because he did not want his wife to think that he was buying lingerie for other women or that he was entertaining himself with the pictures of skimpily clad women in the catalogs. He emphasized that neither he nor his wife had ever purchased from the store—online, in person, or through the catalog. Barton said that since he spent so little money in Washington, he knew that his credit union was the only business having his address. He said that because he believed he should be able to stop financial institutions from selling personal information to third parties, he supported the act. Since the U.S. Congress passed the act, individuals now have the right to direct financial institutions not to sell personal information to third parties (Hoofnagle and Honig, 2005).

In short, if physical access to a computer system can be obtained, U.S. Congress members feared that gaining illegal or inappropriate access to information on that computer could also be obtained. With relatively new U.S. laws pertaining to the security of information on the Internet—including HIPAA (Health Insurance Portability and Accountability Act), the Gramm-Leach-Bliley Act, and the Sarbanes-Oxley Act—data in both physical and electronic forms must not only be protected by adequate access control mechanisms but also be audited, as well, if compliance with the various regulations is going to be maintained.

Trademark Law, Patent Law, and the U.S. Anticybersquatting Consumer Protection Act of 1999

Trademark law governs disputes between business owners over the names, logos, and other means that they use to identify their products and services in the marketplace. More than 33,000,000 Internet domain names have been registered, including tens of thousands of domain names apparently infringing on trademark and service marks. If someone owns a trademark or a service mark (federally registered or not), there could be some domain names infringing on that trademark; although individuals may not realize it, under U.S. trademark law, trademark owners have a duty to police their marks and to prevent other parties from infringing on them (Nolo, 2006).

In 2001, John Zuccarini gained notoriety because of a number of domain name violations of the U.S. Anticybersquatting Consumer Protection Act. By definition, cyber-squatting is registering, trafficking in, or using domain names that are the same as or confusingly similar to existing trademarks—with the "bad faith" intention of profiting from the goodwill of the trademark. Cyber-squatters are like the sooners in pioneer days—people who snatched land before it could be legally claimed and taken as their own (Hampton and McCue, 2005).

On October 30, 2000, the U.S. District Court in Pennsylvania ordered Zuccarini to pay damages of $500,000 (plus more than

$30,000 in attorneys' fees and costs) arising from five Internet domain names he got and used—in violation of the Anticybersquatting Consumer Protection Act. Zuccarini filed an appeal, but the Appeals Court on June 15, 2001, supported the U.S. District Court's decision. Zuccarini ran more than 3,000 Websites, netting himself a profit of between $800,000 to $1,000,000 a year. He registered hundreds of Internet domain names that were misspellings of, but remarkably similar to, famous people's names, marketing brands, company names, actors, television shows, and movies—including Budget Rent a Car Corporation, America Online, Saks and Company, Dow Jones and Company, Nicole Kidman, Minolta, and Microsoft Corporation (Keyt, 2001).

Besides trademark infringement, companies are often highly concerned about patent infringement. In the United States and Canada, patent infringement occurs when an individual makes or sells a patented invention in the said jurisdiction without obtaining authority from the patent owner. Virtually anything "new" can be, arguably, patentable. Recently, patent laws in North America have recognized that software inventions are, for the most part, patentable—as well as methods of doing business. In the United States and in Canada, the limitation period for patent infringement litigation is six years. If a party is found guilty of infringing another's patent, with regard to damages, U.S. patent laws, in particular, say that the court shall award damages adequate to compensate the owner of the patent—such as not less than a reasonable royalty along with interest and costs fixed by the court (Patent Enforcement and Royalties, Ltd., 2002).

Some infringement cases in recent years have resulted in costly legal bills, as lawyers have battled it out in courtrooms over companies' patent rights. One such case was fought over the past four years between EMC Corporation and the Hewlett-Packard Company. Although neither company admitted to any wrongdoing, as part of the settlement finally reached, both companies agreed to a five-year patent cross-licensing agreement, and Hewlett-Packard agreed to pay EMC Corporation $325 million or to buy that amount of EMC Corporation products within five years (In Brief, 2005).

Meanwhile, the patent war between Canadian firm Research in Motion (RIM), Ltd., and the U.S. firm NTP, Inc., continued. Nasty allegations were thrown by one company at the other, despite a recent warning from a U.S. federal judge to settle the patent fight before he enforces a resolution to the five-year-old fight. The judge scolded the executives and lawyers representing the two companies, saying that both should have found a solution by now; he affirmed that if the court had to impose a solution, it would be imperfect. An earlier 2002 jury had found RIM guilty of infringing on five patents belonging to NTP (a Virginia company created to protect the patents of the late high-tech inventor Thomas Campana, Jr.). RIM later lost on appeal as well, and the company failed to convince the U.S. Supreme Court to review the case. (Avery and Waldie, 2006). The case ended in a negotiated settlement in March 2006. RIM's CEO agreed to pay NTP a one-time payment of U.S.$612.5 million. RIM, in return, received a license to NTP's patents going forward.

USA Patriot Act of 2001 and Patriot Act II

On October 27, 2001, the USA Patriot (that is, Uniting and Strengthening America by Providing Appropriate Tools Required to Intercept and Obstruct Terrorism) Act was enacted into law. Congressman Ron Paul told the *Washington Times* that no one in Congress had been permitted to read in full the Patriot Act I text, which was passed very quickly by the House on October 27, 2001. That event, when unveiled, upset many civil libertarians and constitutional scholars. William Safire, writing for the *New York Times*, said that with the passing of Patriot Act I, President Bush was, in effect, seizing dictatorial control.

More heat was added to the debate when, on February 7, 2003, the Center for Public Integrity, a public interest think tank in Washington, D.C., disclosed the entire text (eighty-seven pages) of the Domestic Security Enhancement Act of 2003 (more widely referred to as Patriot Act II), along with a section-by-section analysis of the proposed legislation. The classified document had allegedly been leaked to the Center for Public Integrity by a source inside the federal government whose identity was not made known (Jones, 2006).

Homeland Security Act of 2002

In 2002, the United States passed the Homeland Security Act. Section 225, known as the Cyber Security Enhancement Act of 2002, had the following subsections (U.S. Department of Homeland Security, 2006):

Title I: The Department of Homeland Security (DHS) and its missions and functions.

Title II: Information analysis and infrastructure protection.

Title III: Chemical, biological, radiological, and nuclear countermeasures.

Title IV: Border and transportation security.

Title V: Emergency preparedness and response.

Title VI: Internal management of the DHS.

Title VII: General provisions and coordination with nonfederal entities, the inspector general, and the U.S. Secret Service.

Title VIII: Transitional items.

Title IX: Conforming and other technical amendments.

The U.S. Homeland Security Act of 2002 was brought by Richard Armey to the Standing Committee in the House on July 10, 2002. Amendments were made by the Committee on Homeland Security on July 24, 2002. The legislation passed the House on July 26, 2002, was received in the Senate on November 19, 2002, and was passed by the Senate on November 25, 2002. Called the Homeland Security Act of 2002, it was signed by the president of the United States as Public Law 107–296 to establish the Department of Homeland Security (Center for Democracy and Technology, 2005).

Cyber Security Research and Development Act of 2002

Introduced by Sherwood Boehlert on December 3, 2001, the Cyber Security Research and Development Act was meant to provide money for computer and network security research and for research fellowship programs in the United States. The act was sent to the Committee on Science and to the Committee on Education and the Workforce. On February 7, 2002, the House of Representatives passed the bill. It was read twice before the Senate, was sent to the Committee on Commerce, Science, and Transportation, and became Public Law No. 107–305 (ibid.).

PROTECT Act of 2003, COPPA of 1998, and CIPA of 2000

An important piece of legislation passed in the United States in 2003 to fight Internet-based child pornography was the PROTECT Act—the Prosecutorial Remedies and Tools against the Exploitation of Children Today Act. This act was passed by the U.S. Senate in February 2003, with an extraordinary vote of 84 senators in favor and 0 senators opposed. The objective of this piece of legislation was to assist law enforcement agents in their attempts not only to identify but also to track pedophiles using the Internet to seduce minors. This piece of legislation was seen as a response to the April 16, 2002, Supreme Court decision that overturned most of the CPPA, also known as the Child Pornography Prevention Act of 1966. The case, in point, was *Ashcroft v. Free Speech Coalition* (00–795), 198F.3d 1083.

In other words, the PROTECT Act of 2003 was meant to strengthen the U.S. government's ability to prosecute crimes involving child pornography. The PROTECT Act of 2003 also attempted to extend prosecutorial power beyond U.S. jurisdictions.

The latter implemented the Amber Alert communication system—which allows for nationwide alerts when children go missing or are kidnapped—and redefined child pornography to include not only images of real children engaging in sexually explicit conduct but also computer-generated depictions indistinguishable from real children engaging in such acts. *Indistinguishable* was further defined as that which an ordinary person viewing the image would conclude is a real child engaging in sexually explicit acts. However, cartoons, drawings, paintings, and sculptures depicting minors or adults engaging in sexually explicit acts, as well as depictions of actual adults that look like minors engaging in sexually explicit acts, are excluded from the definition of child pornography (ibid.).

Also, the Children's Online Privacy Protection Act (COPPA) of 1998, effective April 21, 2000, applied to the online collection of personal information from children under the age of thirteen. The rules detailed what a Website operator must include in a privacy policy, when and how to seek verifiable consent from parents or guardians, and what actions an operator must take to protect children's privacy and safety online. It is important to note that these Internet safety policies required the use of filters to protect against access to visual depictions considered obscene or harmful to minors.

A filter is a device or material for suppressing or minimizing waves or oscillations of certain frequencies. Therefore, filtering software should block access to Internet Websites listed in an internal database of the product, block access to Internet Websites listed in a database maintained external to the product itself, block access to Internet Websites carrying certain ratings assigned to those Websites by a third party or that are unrated under such a system, and block access based on the presence of certain words or phrases on those Websites. In short, software filters use an algorithm to test for appropriateness of Internet material—in this case, for minors. Websites are first filtered based on IP addresses or domain names. Since that process is based on predefined lists of appropriate and inappropriate Websites, relying totally on such lists is ineffective because Internet Websites come and go so quickly. Moreover, though minors often frequent online chat rooms, instant messaging, and newsgroups, these are not under the filtering system.

Under the Children's Internet Protection Act (CIPA) of 2000, U.S. libraries receiving e-rate discounts, or grants for Internet access, are required to enforce policies using technological protection measures—particularly filters—to prevent minors from viewing visual depictions falling into three categories: child pornography, obscenity, and information that is generally harmful to minors. However, this law does not require filters to block other text-based material (Minow, 2003).

Royal Canadian Mounted Police (RCMP) corporal Jim Gillis, head of Project Horizon, a policing initiative dealing with online child pornography and based in Halifax, Nova Scotia, Canada, maintains that the Internet pornography industry generates a whopping $57 billion annually worldwide and supports more than 4 million Websites. In that group, notes Jim Gillis, there are more than 100,000 child pornography sites creating about $2.5 billion a year in revenues (Butters, 2005; Mahoney, 2005).

CAN-SPAM Act of 2003

The CAN-SPAM Act was passed by the U.S. Senate on November 25, 2003, with the purpose of doing what its name implies: Controlling the Assault of Non-Solicited Pornography and Marketing Act of 2003.

The objective was to target, in particular, spammers and massive e-mailers whose objectives—according to the e-mail headers—appeared to be legitimate but were not. The reason that Congress made this daring move was that, according to 2003 estimates, unsolicited commercial e-mail was estimated to account for more than half of all e-mail traffic, a rise from an estimated 7 percent just two years earlier. Moreover, the U.S. Congress affirmed that the volume of spam continues to rise—and that most of these electronic messages are fraudulent or deceptive.

Under the CAN-SPAM Act, chapter 47 of title 18, U.S. Code, was amended at the end of new section 1037, Fraud and Related Activity in Connection with Electronic Mail. Offenders of the act are considered to be those individuals in or affecting interstate or foreign commerce who knowingly:

- Gain access to a protected computer without authorization and intentionally initiate the sending of multiple commercial e-mail messages from or through that computer and a connection to the Internet.
- Use a protected computer to relay or retransmit multiple commercial e-mail messages to either intentionally deceive or to mislead receivers or any Internet access service as to the origin of such e-mail messages.
- Intentionally falsify header information in multiple commercial e-mail messages and intentionally initiate the sending of such messages.
- Register using information that falsifies the identity of the actual registrant for five or more e-mail accounts, or online user accounts of two or more domain names, and intentionally initiate the sending of multiple commercial e-mails from any combination of these accounts or domain names.
- Intentionally falsely represent oneself as a legitimate registrant of five or more Internet Protocol (IP) addresses, and then intentionally initiate the sending of many commercial e-mail messages from such addresses.

Those breaching this act receive a fine, imprisonment for not more than five years, or both. The act is enforced by the U.S. Federal Trade Commission (FTC), and within six months of its enactment, the FTC sent to the Senate Committee on Commerce, Science and Transportation as well as to the House of Representatives Committee on Energy and Commerce a report detailing a plan and a timetable for establishing a U.S.-wide marketing Do-Not-E-Mail registry (Sorkin, 2006).

Provisions of many of the following state laws regarding spam may be pre-empted by the CAN-SPAM Act of 2003: Alabama, Alaska, Arizona, California, Colorado, Connecticut, Florida, Georgia, Hawaii, Idaho, Illinois, Kansas, Kentucky, Louisiana, New Jersey, New York, Pennsylvania, and Rhode Island—to name just a few.

Other jurisdictions around the globe have passed similar antispam laws. For example, the European Union (EU) has a number of pieces of legislation, including: the E-Privacy Directive, the E-Commerce Directive, and the Data Protection Directive (ibid.)

Moreover, Canada's anti-intrusion legislation involves section 342.1 of the Canadian Criminal Code, aimed at several potential harms to property and persons, including stealing computer services, invading individuals' privacy, and trading in computers' passwords or cracking encryption systems. Charges for such breaches are usually made with regard to applicable sections of the Criminal Code—theft, fraud, computer abuse, information abuse, and the interception of communications.

Cyber Security Enhancement Act of 2005/2001

The Cyber Security Enhancement Act of 2001 was introduced and sent to the House Judiciary by Lamar Smith on December 13, 2001, to provide for greater cyber-security for the United States. A hearing was held in the Crime Subcommittee on February 26, 2002. On July 16, 2002, it was sent to the Senate committee, read two times, and then sent to the Committee on the Judiciary.

On April 20, 2005, the House Homeland Security Subcommittee on Economic Security, Infrastructure Protection, and Cybersecurity passed HR 285, the Cyber Security Enhancement Act of 2005. The act stated not only that the assistant secretary for cyber-security would be the head of the Directorate's National

Cyber Security Division but also that the divisions would identify and reduce vulnerabilities and threats, as well as create cyber-attack warning systems (Dizzard III, 2005).

A Sampling of Recent Internet Crime Cases

A number of interesting Internet crime cases have occurred over the past five years. All of these are press releases listed under the U.S. Department of Justice Website, http://www.usdoj.gov/ (which is in the public domain), and pertain to individuals charged and convicted under the U.S. Computer Crime Statute, U.S.C. section 1030.

Before detailing ten of these court cases—meant to be a representative rather than a comprehensive listing of the Internet legal cases detailed on this Department of Justice Website over the past five years—it is important to note that, as of May 23, 2001, the FBI and the Department of Justice announced a nationwide sweep into Internet fraud. The project, called Operation Cyber Loss, was started by the FBI's Internet Fraud Complaint Center (IFCC). The project has been coordinated by FBI offices, the U.S. Postal Inspection Service (USPIS), the Internal Revenue Service (Criminal Investigative Division), U.S. Customs Service, the U.S. Secret Service, and many other state and local law enforcement arms (Kubic, 2001).

The Internet fraud schemes exposed as part of this investigation have represented more than 56,000 online users who have claimed cumulative losses exceeding $117 million. The schemes highlighted by the project have included online auction fraud, the regular nondelivery of merchandise bought over the Net, debit and credit card fraud, identity theft, security fraud, and multilevel marketing or pyramid schemes. As a result of this project, more than ninety Internet crime perpetrators have been charged with a variety of crimes, including wire fraud, mail fraud, conspiracy to commit fraud, bank fraud, money laundering, and Intellectual Property Rights infringement. As is typical of Internet fraud, the perpetrators and the targets involved in this operation have been scattered throughout the world, with many of the schemes being multijurisdictional in nature. Moreover, the cases tend to reflect the propensities of those committing fraud:

they tend to go from one fraud scheme to another trying to "rip off" others for their own personal gain. The only way to stop them in their tracks is to charge them and convict them (ibid.).

2002

Russian Computer Hacker Sentenced to Three Years in Prison (October 4) *[United States v. Gorshkov]*

At age twenty-seven, Vasiliy Gorshkov, a Russian, was sentenced to three years in jail for charges including network cracking, conspiracy, and fraud committed against the Speakeasy Network of Seattle, Washington; the Nara Bank of Los Angeles, California; and the online credit card payment company PayPal of Palo Alto, California. Gorshkov also had to pay nearly $700,000 in restitution for losses he caused to Speakeasy and PayPal.

Vasiliy was one of two men from Russia (the other being Alexy Ivanov, age twenty-three) conned into traveling to the United States as part of an FBI undercover operation to get the individuals responsible for these crimes. A nationwide investigation was started by the FBI into Russian computer intrusions directed at U.S. Internet service providers, e-commerce Websites, and online banks. The FBI was able to determine that the crackers gained unauthorized access to the targeted computers to steal credit card information and other personal financial information. The perpetrators also often tried to extort money from the targets with threats of exposing sensitive data to the public, or of damaging the targeted computers. The crackers also defrauded PayPal through a plot in which stolen credit cards were used to generate money and to pay for computer parts purchased from companies in the United States.

Creator of Melissa Computer Virus Sentenced to Twenty Months in Federal Prison (May 1) *[United States v. Smith]*

The creator of the Melissa computer virus, David L. Smith, age thirty-four, of New Jersey, pleaded guilty in court of developing the destructive virus. The accused said he recognized that his Melissa virus had caused more than $80 million in damages because of disruptions to computer networks in businesses and government offices. The judge hearing the case ordered Smith to serve three years of supervised release after he finished serving a twenty-month prison sentence. Smith was also fined $5,000 and was told that he could not be involved with computer networks, the Internet, or Internet bulletin board systems unless authorized by the court. He also had to serve 100 hours of community service in a way that would best make use of his technology know-how.

2003

Disgruntled Philadelphia Phillies Fan Charged with Hacking into Computers Triggering Spam E-Mail Attacks (October 7) *[United States v. Carlson]*

An interesting court case occurred in the fall of 2000. A disgruntled Philadelphia Phillies baseball fan by the name of Allan Eric Carlson was arrested at his home and charged by FBI agents with cracking into computers around the United States and using them as launch sites for his cyber exploits in 2001 and 2002. He hijacked or spoofed the return addresses of reporters at the *Philadelphia Inquirer* newspaper, the *Philadelphia Daily News*, and the Philadelphia Phillies' offices and then used those e-mail addresses to launch spam e-mail attacks. The FBI agents also charged Carlson with identity theft for illegally utilizing the e-mail addresses of the newspaper reporters.

The judge hearing the case said that fans have the right to voice their displeasure, but these acts committed by Carlson were electronic attacks with serious consequences, for by flooding the targeted computer systems with spam e-mails, those systems as well as the businesses in which they were placed were seriously harmed. The judge affirmed that disgruntled fans can boo and they can turn off the television, but they cannot hijack the e-mail addresses of unsuspecting users and call it a passion. The judge also affirmed that this was the first use of an identity theft statute against an e-mail spammer. The accused faced a maximum possible sentence of 471 years' imprisonment, $117,250,000 in fines, and a special assessment of $7,800.

Los Angeles, California, Man Sentenced to Prison for Role in International Computer Hacking and Internet Fraud Scheme (February 28) *[United States v. Pae]*

In February 2003, a twenty-year-old named Thomas Pae was sentenced in the United States on charges of wire fraud, credit card fraud, and conspiracy. He was a partner in a seven-person international computer cracking and Internet fraud scheme. The target was Ingram Micro, a wholesale distributor of technology products based in Santa Ana, California.

Pae pleaded guilty to the charges and admitted that in 2001 he had partaken in a scheme with others in Romania to gain unauthorized access to Ingram Micro's online ordering system to place hundreds of thousands of dollars worth of computer equipment orders. The equipment was then sent to destinations controlled by Pae and other team players in the Los Angeles area. This equipment was then repackaged and sent to Eastern Europe, where it was likely resold. During his hearing, Pae also said that he had purchased credit card numbers from crackers on the Internet and used them to buy computer chips, hard drives, PDAs, and other tech items from e-commerce Websites like Handspring, Amazon, and Egghead. He also said that he and his team members had tried to buy more than $500,000 in computer equipment and other items from Ingram Micro and the other on-line retailers.

In the end, Pae was ordered to pay $324,061 in restitution to the targets and to spend thirty-three months behind bars. After his release, Pae was to serve three years of supervised release.

2004

U.S. Charges Hacker with Illegally Accessing *New York Times* Computer Network (January 8) *[United States v. Lamo]*

Adrian Lamo, age twenty-two, a friend of infamous cracker Kevin Mitnick, was charged in a New York court with cracking into the *New York Times* newspaper computer network in 2002 and accessing a database having personal information (such as the home telephone numbers and Social Security numbers) of more than 3,000 contributors to the op-ed page of the paper.

Lamo added an entry to the database for himself; he listed his personal information (such as his cell phone number—which was 415-505-HACK) and a summary of his areas of expertise: "computer hacking, national security, communications intelligence." While inside the internal network, he created five fake user ID names and passwords for the *New York Times*'s account with LexisNexis, an Internet-based subscription service providing legal, news, and other useful information. Then, over a three-month period, Lamo used these five fictitious user IDs and passwords to complete more than 3,000 searches on LexisNexis, at a total cost of about $300,000. After pleading guilty to his cyber exploits, Lamo received six months' detention at his parents' home, was placed on two years' probation after his completion of that term, and was ordered to pay $65,000 in restitution.

Justice Department Sues to Halt Alleged Internet Tax Scam (September 9) *[United States v. Luman]*

The U.S. Justice Department requested on September 9, 2004, that a federal court in Georgia bar Jonathan Luman from selling a tax scam through the Internet. According to the civil injunction complaint, this individual operated Websites selling fake documents as well as a "Tax Buster Guide"—with forms that purchasers could submit to the Internal Revenue Service instead of the usual tax return forms.

Luman informed potential customers that his documents would let them reclaim their "sovereign citizen" status and eliminate liability for federal income taxes. His guide sold for $50. The guide and completed forms were sold for about $200 to customers in more than forty U.S. states. The Justice Department also wanted an order directing the perpetrator to give them any records identifying his Internet clients and an order to require him to post the court order on his Websites. The assistant attorney general for the Justice Department's Tax Division said that illegal tax scams utilizing the Internet are costing honest taxpayers billions of dollars annually. Such scams are listed on the Internal Revenue Service's "Dirty Dozen" list, at http://www.irs.gov/newsroom/article/0,id=120803,00.html.

2005

Internet Pharmacy Operator Receives 51-Month Prison Sentence (January 21) *[United States v. Kolowich]*

A U.S. district court judge sentenced Mark Kolowich to federal prison for fifty-one months and repayment of a large amount of cash proceeds for operating one of the largest Internet pharmacy schemes ever prosecuted in the United States. His online company was known as World Express RX.

Kolowich pleaded guilty to selling counterfeit pharmaceuticals over the Internet, committing mail fraud, smuggling pharmaceuticals, and conspiring to launder money. In a related case, he also pleaded guilty to importing unapproved drugs into the United States, introducing those unapproved drugs to interstate commerce, and smuggling unapproved drugs into the country. The attorney for the Department of Justice said that he should get criminal penalties for diverting and counterfeiting prescription drugs for profit. When consumers use the Net to get prescription drugs, they take a risk that the drugs sold to them may not be what they appear to be.

The perpetrator ran the pharmacy Website at www.WorldExpressRx.com, and consumers could buy prescription drugs without a prescription. The Website told potential customers simply to complete a health questionnaire and pay a $35 flat fee for a doctor's consultation. In fact, the Website falsely represented that a physician would review the questionnaire and afterward issue a lawful prescription before the drugs would be shipped. Kolowich had no physician affiliated with this scheme. The distributed drugs contained active ingredients for such prescription drugs as Viagra, Cialis, Levitra, Propecia, Celebrex, and Xenical. Apparently, the drugs were made in Mexico and smuggled into California. They were then packaged and sent to online customers throughout the United States and elsewhere around the globe. Some of the ingredients were imported from China and India with false shipping documents.

This case was investigated by the Federal Bureau of Investigation, the Food and Drug Administration Office of Criminal Investigations, the Internal Revenue Service Criminal Investigations Division, the Immigration and Customs Enforcement Division of the Department of Homeland Security, the U.S. Postal Inspection Service, and the Chula Vista Police Department.

First Criminal Defendants Plead Guilty in Peer-to-Peer Copyright Piracy Crackdown (January 18) *[United States v. Trobridge and Chicoine]*

In a U.S. courtroom, William Trowbridge, age fifty, and Michael Chicoine, age forty-seven, of Texas both pleaded guilty to conspiracy and committing felony copyright infringement. This was the first federal conviction in the United States for copyright piracy using peer-to-peer (P2P) networks.

The attorney for the Department of Justice said that the theft of Intellectual Property victimizes the owners of the creative works as well as the American people—who are then forced to shoulder the burden of increased costs for such goods and services. The pair were caught through an FBI operation known as Operation Digital Gridlock, targeted at illegal file-sharing of copyrighted information over five Direct Connect P2P networks belonging to an online group of hubs known as The Underground Network. These networks operated by having users share large quantities of computer files with other Internet users, all of whom could download each other's files.

For two years Trowbridge owned and operated a Direct Connect hub called the Movie ®oom, and Chicoine owned and operated a hub named Ãçhènøñ's Alley™ (pronounced "Achenon's Alley"). The accused offered a wide variety of computer software, computer games, music, and movies in digital format, including some software titles selling on the normal market for thousands of dollars.

During the investigation, government agents downloaded thirty-five copyrighted works worth about $5,000 from Chicoine's hub, and more than seventy copyrighted works worth about $21,000 from Trowbridge's hub. The pair pleaded guilty. The maximum penalties for a first-time offender convicted of conspiracy to commit felony criminal copyright infringement in violation of Title 17, U.S. Code, Section 506, and Title 18, U.S. Code, Sections 371 and 2319, are five years in prison, a fine of $250,000, restitution paid to the targets, and the release to authorities and destruction of infringing copies and all equipment used to manufacture such illegal copies.

2006

Second Defendant Pleads Guilty in Prosecution of Major International Spam Operation (January 31) [United States v. Rogers]

A California man named Kirk Rogers, age forty-three, pleaded guilty to aiding and abetting others to violate the CAN-SPAM Act of 2003. This was the California man's second conviction related to the sending of obscene e-mails through the Internet. Rogers agreed to forfeit money obtained by those crimes, and he faces a maximum sentence of five years in prison if found guilty. His sentencing will not take place until after June 2006.

According to the plea agreement, he helped three people send spam e-mails containing graphic pornographic images. Rogers developed and managed the computer system used to send spam e-mail on behalf of two of his partners. The attorney representing the Justice Department said that this prosecution addresses two of the most significant problems arising with the

Internet: online obscenity and the intrusive, costly, and often harmful practice of e-mail spam. The prosecution of this case would also help stop pornographers wanting to enter houses uninvited and threaten children's innocence. This case, the first in the country, should send a strong message to other pornographers, affirmed the attorney.

Apparently America Online, Inc., received more than 600,000 complaints between January 30, 2004, and June 9, 2004, from its online users regarding spam e-mails that had allegedly been sent by this spamming operation. The indictment further alleged that the spam e-mails sent by the accused marketed pornographic Internet Websites to earn commissions for directing Internet traffic to these Websites. It was also alleged that graphic pornographic images were embedded in each of the defendants' e-mails. Four counts of the indictment charged felony obscenity offenses for the sending of hard-core pornographic images of adults engaged in explicit sexual conduct, meeting the Supreme Court's test for adult obscenity.

According to the indictment, the spam e-mails were sent in a way that would impair the ability of recipients, Internet service providers processing the e-mails on behalf of recipients, and law enforcement agencies to identify, locate, or respond to the senders. The indictment further alleged that Rogers's two partners in crime created and used overseas companies named the Compliance Company and Ganymede Marketing to hide their exploits. His two partners in Internet crime also used overseas bank accounts in Mauritius and the Isle of Man to launder and distribute proceeds from the spamming exploits.

Virginia Man Pleads Guilty to Child Pornography Charges (January 27) [United States v. Mitchel]

A Virginia man by the name of Gregory Mitchel, age thirty-eight, pleaded guilty to multiple charges involving the sexual exploitation of minor boys and operating child pornography Websites on the Internet. He was charged with four counts, including the production and selling of child porn through the Net through child pornography Websites. In court, Mitchel pleaded guilty to the production, distribution, and sale of child pornography.

An investigation by the FBI ended in his arrest on September 12, 2005. Mitchel operated a Website called Justinsfriends.com— which sold membership subscriptions to online users wanting to get videos of minor boys engaging in sexually explicit conduct.

Mitchel was also responsible for producing content for the Web-sites by filming videos of minors engaging in explicit sex acts. Mitchel and other partners then received the proceeds from membership subscriptions.

Because of a previous child pornography conviction, Mitchel faces a mandatory minimum sentence of not less than twenty-five years to a maximum of fifty years for both the production and dis-tribution of child pornography. The charge of selling child pornog-raphy carries a sentence of not less than fifteen to forty years in prison. Finally, the charge of possession of child pornography car-ries a sentence of not less than ten years to a maximum of twenty years. The accused also faces lifetime supervised release if he is ever released from prison. The United States is also seeking forfei-ture of the proceeds used in committing these Internet exploits.

Data Regarding Internet Use and Prevalence Worldwide

TABLE 6.1: Top 15 Internet-Using Countries

Year-end 2004:	Internet Users (#K)	Share %
1. United States	<186,000	<20
2. China	<100,000	<11
3. Japan	<79,000	<9
4. Germany	<42,000	<5
5. India	<37,000	<4
6. United Kingdom	<34,000	<4
7. South Korea	<32,000	<4
8. Italy	<26,000	<3
9. France	<26,000	<3
10. Brazil	<23,000	<3
11. Russia	<22,000	<3
12. Canada	<21,000	<3
13. Mexico	<14,000	<2
14. Spain	<14,000	<2
15. Australia	<14,000	<2
Top 15 Countries	*<663,000*	*<71*
Worldwide Total	**<935,000**	

Source: Computer Industry Almanac, Inc., 2004.

TABLE 6.2:
Selected 2006 Internet2 Connectivity Survey Findings
Regarding Internet Connectivity in U.S. Educational Environments

State Network Overview:

- About 35 networks are connected to the high-performance Internet2 backbone.
- The average State Education Network connects to Internet2 at 756 Mbps, up from 467 Mbps in 2004 and from 138 Mbps in 2002.
- Almost 70% of the state education networks connected to the Internet2 backbone network are multicast-enabled.
- About 60% of the state education networks accessing the Internet2 backbone network provide other video service, such as MPEG.

Four-Year Colleges and Universities:

- About 1,002 of 2,026, or 49%, of the four-year colleges and universities in the United States are connected to Internet2.
- About 69 of these 1,002, or 57%, of the four-year colleges and universities in the United States connect to the Internet2 backbone network at or more than 10 Mbps.

Community Colleges:

- About 677 of 1,252, or 54%, of the community colleges in the United States are connected to Internet2.
- About 328 of 677, or 48%, of community colleges connect to the Internet2 backbone network at or more than 10 Mbps.
- About 17% of state education networks report that between half to all of the community colleges they connect are multicast-enabled.

Elementary Schools:

- About 35,971 of 98,335, or 37%, of the elementary schools in the United States are connected to Internet2.
- About 4,350 of 35,971, or 12%, of elementary schools connect to the Internet2 backbone network at or more than 10 Mbps.
- About 17% of state education networks report that between half to all of the elementary schools they connect are multicast-enabled.
- State education networks report that, on average, 44% of the elementary schools they connect have MPEG or other video conferencing equipment available.

Public Libraries:

- About 3,325 of 16,991, or 20%, of the public libraries in the United States are connected to Internet2.
- About 229 of 3,325, or 7%, of public libraries connect to the Internet2 backbone network at or more than 10 Mbps.

TABLE 6.2: Continued

Museums, Historical, and Cultural Centers:
- About 28 of 80, or 35%, of museums, historical, and cultural centers connect to the Internet2 backbone network at or more than 10 Mbps.
- About 19 of 24, or 79%, of performing arts centers connect to the Internet2 backbone network at or more than 10 Mbps.

Science Centers, Planetariums, and Observatories:
- About 36 of 228, or 16%, of the science centers, planetariums, and observatories in the United States are connected to Internet2.
- About 25 of 36, or 68%, of science centers, planetariums, and observatories connect to the Internet2 backbone network at or more than 10 Mbps.

Zoos and Aquariums:
- About 12 of 203, or 6%, of the zoos and aquariums in the United States are connected to Internet2.
- About 2 of 12, or 17%, of zoos and aquariums connect to the Internet2 backbone network at or more than 10 Mbps.

Source: Rotman, Lauren. "Internet Survey Finds over 46,000 K20 Institutions Connected to Its Next-generation Network." Lauren Rotman, https://mail.internet2.edu/wws/arc/i2-news/2006–03/msg00004.html (accessed March 24, 2006).

TABLE 6.3: Top 15 Internet-Using Countries in Asia

Year-end 2005:	Internet Users	Share %
1. China	111,000,000	<31
2. Japan	86,050,000	<24
3. India	50,600,000	<14
4. South Korea	33,900,000	<10
5. Indonesia	18,000,000	<5
6. Taiwan	13,800,000	<4
7. Malaysia	10,040,000	<3
8. Thailand	8,420,000	<3
9. Philippines	7,820,000	<3
10. Pakistan	7,500,000	<3
11. Vietnam	5,870,000	<2
12. Hong Kong	4,878,713	<2
13. Singapore	2,421,000	<1
14. Uzbekistan	880,000	<1
15. Azerbaijan	408,000	<1
Total Asia	*364,270,713*	

Source: www.Internetworldstats.com, 2006.

TABLE 6.4: Internet Pornography Statistics

Pornography Industry Revenue Statistics:
  Size of the Industry: $57 billion worldwide
  Adult Videos: $20 billion
  Internet Materials: $2.5 billion
  Child Pornography: $3 billion

Internet Pornography Statistics:
  Pornographic Websites: 4.2 million
  Pornographic Pages: 372 million
  Daily Pornographic Search Engine Requests: 68 million
  Monthly Pornographic Downloads/user: 4.5
  Websites Offering Illegal Child Porn: 100,000
  Sexual Solicitation of Minors in Chat Rooms: 89%

Children Internet Pornography Statistics:
  Average Age of Exposure to Internet Porn: 11 years
  Largest Consumers of Internet Porn: 12-17 years
  Percentage of 8- to 16-year-olds Viewing Online Porn: 90%

Adult Internet Pornography Statistics:
  U.S. Adults Who Regularly Visit Online Porn Sites: 40 million
  Breakdown of Male-Female Visitors to Online Porn Sites: 72% Male, 28% Female
  Adults Admitting to Online Sexual Addiction: 10%

Source: Ropelato, Jerry. "Internet Pornography Statistics." Internet Filter Review, http://internet-filter-review.toptenreviews.com/internet-pornography-statistics.html (accessed 2005).

References

Avery, S., and P. Waldie. "RIM, NTP Escalate Their War of Words." *Globe and Mail,* February 28, 2006, p. B5.

Butters, George. "Criminal Activity: Your Computer May Be Housing Child Porn," *Globe and Mail,* January 27, 2005, p. B14.

Center for Democracy and Technology. "Legislation Affecting the Internet." Center for Democracy and Technology, http://www.cdt.org/legislation/107th/wiretaps/ (accessed 2005).

Centers for Medicaid and Medicare Services. "The Health Insurance Portability and Accountability Act of 1996 (HIPAA)." Centers for Medicaid and Medicare Services, http://www.cms.hhs.gov/hipaa/ (accessed October 16, 2002).

Computer Industry Almanac, Inc. "Worldwide Internet Users Will Top One Billion in 2005." Computer Industry Almanac, http://www.c-i-a.com/pr0904.htm (accessed September 3, 2004).

Dizzard III, Wilson. "Bill to Promote Cyber Security Chief Moves Forward." Post-Newsweek Media, Inc., http://www.gcn.com/vol1_no1/daily-updates/35577-1.html (accessed April 20, 2005).

Electronic Privacy Information Center. "The Gramm-Leach-Bliley Act." Electronic Privacy Information Center, http://www.epic.org/privacy/glba/ (accessed January 21, 2005).

Hampton, P., and J. McCue. "Internet Corporation for Assigned Names and Numbers/Anticybersquatting Consumer Protection Act of 1999: Dealing with Cyber Claims." www.dicksteinshapiro.com/files/Publication/c752393edc1c–4562–842b–2ef4ee175678/Presentation/PublicationAttachment/784e1910–3c69–4d54–9074–3344e6e4f471/ALA%20class%20Journal,%20Internet%20Corporation%20For%20Assigned%20Names,%2012–1.pdf (accessed December 2005).

Hoofnagle, C., and E. Honig. "Victoria's Secret and Financial Privacy." Epic, http://www.epic.org/privacy/glba/victoriassecret.html (accessed January 25, 2005).

In Brief. "HP and EMC Settle Patent Infringement Case." *Globe and Mail,* May 5, 2005, p. B25.

Jones, Alex. "Secret Patriot Act II Destroys Remaining US Liberty." Rense.com, http://www.rense.com/general34/takeover.htm (accessed February 25, 2006).

Keyt, Richard. "Notorious Cybersquatter Liable for $500,000 under the Anticybersquatting Consumer Protection Act." Keyt Law, http://www.keytlaw.com/urls/zuccarini.htm (accessed June 30, 2001).

Kubic, Thomas. "Congressional Testimony." U.S. Department of Justice, http://www.fbi.gov/congress/congress01/kubic052301.htm (accessed May 23, 2001).

Mahoney, Jill. "Child-porn Charges Up, Statistics Canada Says." *Globe and Mail,* April 21, 2005, p. A6.

Minow, Mary. "Children's Internet Protection Act (CIPA): Legal Definitions of Child Pornography, Obscenity and 'Harmful to Minors.'" Law Library Resource Exchange, http://www.llrx.com/features/updatecipa.htm. (accessed August 31, 2003).

Nolo, Inc. "Trademarks and Copyright." Nolo, Inc., http://www.nolo.com/lawcenter/ency/index.cfm/catID/804B85E3–9224–47A9–7E6B5BD92AACD48 (accessed March 1, 2006).

Patent Enforcement and Royalties, Ltd. (Pearl). "All About Patents: What Is Intellectual Property?" Pearl, Ltd., http://www.pearlltd.com/content/all_about_patents.html (accessed February 13, 2002).

Ropelato, Jerry. "Internet Pornography Statistics." Internet Filter Review, http://internet-filter-review.toptenreviews.com/internet-pornography-statistics.html (accessed 2005).

Rotman, Lauren. "Internet Survey Finds over 46,000 K20 Institutions Connected to Its Next-generation Network." Lauren Rotman, https://mail.internet2.edu/wws/arc/i2-news/2006–03/msg00004.html (accessed March 24, 2006).

Sorkin, David. "Spam Laws: United States: Federal Laws." SpamLaws.com, http://www.spamlaws.com/federal/index.shtml (accessed February 27, 2006).

U.S. Department of Homeland Security. "Homeland Security Act of 2002." U.S. Government, http://www.whitehouse.gov/deptofhomeland/analysis (accessed March 1, 2006).

U.S. Department of Justice. "Press Releases." U.S. Department of Justice, http://www.usdoj.gov (accessed March 1, 2006).

www.Internetworldstats.com. "Internet Usage in Asia." Internet World Stats, http://www.internetworldstats.com/stats3.htm#asia (accessed April 9, 2006).

7

Directory of Organizations

T his chapter gives readers an overview of the agencies and organizations related to Internet standards development and
maintenance, as well as those safeguarding privacy, security,
and trust issues. The first part of the chapter presents government and government-affiliated agencies having those objectives. The next part lists independent organizations, and the last
part gives educational institutions offering degrees in Internet
safety and security.

Government Agencies

Federal Bureau of Investigation (FBI)
J. Edgar Hoover Building
935 Pennsylvania Ave. NW
Washington, DC 20535–001
Tel: (202) 324–2000
Fax: (202) 324–3367
Web: www.fbi.gov

The Federal Bureau of Investigation (FBI) Website, besides providing updates on the "most wanted" criminals and terrorists
who have been caught working with partners around the globe,
has listed these priorities: counterterrorism, counterintelligence,
cyber-crime and Internet crime cases, public corruption, civil
rights issues, organized crime, and white collar crime.

Federal Trade Commission (FTC)
600 Pennsylvania Ave. NW
Washington, DC 20580
Tel: Consumer Response Center: (877) 382–4357
Identity Theft Complaint Center: (877) 438–4338
Fax: (202) 326–2034
Web: www.ftc.gov

The Federal Trade Commission (FTC), headquartered in Washington, D.C., developed a national spam database in 2003. The FTC asked e-mail users disgusted with spam to forward to them spam messages so that the FTC could better track the problem nationally. The FTC affirmed that in one year, alone, they received more than 17 million complaints about spam, and they received almost 110,000 complaints daily. Prior to the CAN-SPAM Act's passage, on April 17, 2003, the FTC asked an Illinois judge to block a spam operation of Brian Westby, who used a combination of bland subject lines, fake return addresses, and false "reply to" links to con naive clients to visit Websites offering pornographic material.

Designating March as Fraud Prevention Month, the FTC encourages Internet users to contact them to report incidents of online fraud, spam, and identity theft.

Internet Crime Complaint Center (IC3)
Web: www.ic3.gov/

Internet fraud includes a wide range of online criminal activities that harm targets. Internet fraud encompasses credit card fraud, online auction fraud, unsolicited e-mail (spam) fraud, and online child pornography. In the United States, the Internet Crime Complaint Center (IC3), a partnership between the FBI and the National White Collar Crime Center (NW3C), was created to address Internet fraud.

IC3's mission is to act as a channel to receive, develop, and refer criminal complaints regarding Internet crime to the FBI and the National White Collar Crime Center. The IC3 gives targets a user-friendly way of alerting authorities of suspected criminal or civil violations through the IC3 Website. For law enforcement and regulatory agencies at the federal, state, local, and international levels, IC3 provides a central referral service for complaints regarding Internet abuses.

National Infrastructure Protection Center (NIPC)
Information Analysis Infrastructure Protection
Washington, DC 20528
Tel: (202) 323–3205
Fax: (202) 323–2079
Web: www.nipc.gov/

The NIPC is operated by the Federal Bureau of Investigation (FBI), in cooperation with the U.S. Department of Homeland Security. Besides the NIPC, the FBI has also set up a number of Internet abuse units across the United States. According to the NIPC Website, the NIPC acts as a critical infrastructure threat assessment unit, a warning unit, a threat vulnerability unit, and a law enforcement investigation and response unit. The NIPC provides timely warnings of international threats and gives details to law enforcement about how best to respond to homeland security threats.

National Security Agency (NSA)
Tel: (301) 688–6524
Web: www.nsa.gov/home_html.cfm

The National Security Agency/Central Security Service is the key cryptologic organization in the United States, whose function it is to coordinate, direct, and conduct specialized activities to safeguard the government's IT networks and to produce foreign signals intelligence data. A high-tech organization, the NSA is continuously on top of the evolving frontier of Internet communications and data processing. The NSA also acts as an important center for foreign-language analysis and research.

Office of Critical Infrastructure Protection and Emergency Preparedness (OCIPEP) and Public Safety and Emergency Preparedness Canada (PSEPC)
340 Laurier Ave. West
Ottawa, Canada K1A 0P8
Tel: (613) 991–3283
Fax: (613) 998–9589
Web: http://www.psepc-spcc.gc.ca/index-en.asp

Canadian prime minister Jean Chrétien announced on February 5, 2001, the development of OCIPEP. The OCIPEP (Office of Critical Infrastructure Protection and Emergency Preparedness) took over the functions of the former Emergency Preparedness

Canada. The role of OCIPEP was to protect Canada's critical infrastructures from disruption or complete failure and to ensure the health, safety, and economic well-being of Canadians. Without question, a prolonged disruption or failure in one utility contributing to the critical infrastructure could produce cascading disruptions or failures across a number of other infrastructures, with major negative economic and social repercussions for Canadians.

In December 2003, Canadian prime minister Paul Martin said that OCIPEP would be integrated into a new department known as Public Safety and Emergency Preparedness Canada (known as PSEPC).

U.S.-Computer Emergency Response Team (US-CERT)
Web: www.us-cert.gov

US-CERT is a partnership between the U.S. Department of Homeland Security and the public and private sectors. Established in 2003 to protect the nation's Internet infrastructure, US-CERT analyzes and reduces cyber-threats and vulnerabilities, disseminates cyber-threat warning information across the nation, and coordinates incident responses. The system is used to report cyber-related incidents to US-CERT. "Incidents" include the following:

- attempts (failed or successful) to gain unauthorized access to a computer system or its data.
- undesired disruptions or denial of service to authorized Internet users.
- unauthorized use of a computer system for the processing or storage of data.
- changes to computer system hardware, firmware, or software characteristics without the owner's knowledge, instruction, or consent.

Incidents should be reported to https://forms.us-cert.gov/report/. Phishing e-mail should be sent to phishing-report@us-cert.gov.

U.S. Department of Homeland Security
Homeland Security Operations Center
Washington, DC 20528

Tel: (202) 282–8101
E-mail: HSCenter@dhs.gov
Web: www.dhs.gov

After the September 11, 2001, terrorist attacks on the World Trade Center and the Pentagon, President George W. Bush decided that twenty-two previously disparate domestic agencies should be co-ordinated into one department to provide for better homeland protection and to prevent future terrorist attacks. Thus, the Department of Homeland Security was created.

Component agencies assist in a variety of ways, including analyzing threats and intelligence, guarding surrounding borders and airports, protecting critical infrastructures, and coordinating responses to future emergencies. Besides providing increased security to the U.S. homeland, the department is also committed to protecting the rights of U.S. citizens and enhancing public services (such as natural disaster assistance—evident following hurricane Katrina) by dedicating offices to those important missions.

A six-point agenda for the Department of Homeland Security was recently updated and includes the need to:

- Increase overall preparedness, especially for events deemed to be catastrophic.
- Create improved transportation security systems to transport both persons and cargo more efficiently and safely.
- Strengthen and improve border security and reform immigration processes.
- Enhance intelligence information-sharing with partnering nations, including through the Internet.
- Improve human resource development for members of the Department of Homeland Security, promote and follow sound fiscal management, and optimize procurement and Information Technology issues (including the use of the Internet).
- Realign the Department of Homeland Security's organization so as to maximize mission objectives and performance.

U.S. Department of Justice
10th and Constitution Ave. NW
Criminal Division (Computer Crime and Intellectual Property
    Section)

John C. Keeney Building, Suite 600
Washington, DC 20530
Tel: (202) 514–1026
Fax: (202) 514–3546
Web: www.cybercrime.gov

The U.S. Department of Justice (whose general Website is found at www.usdoj.gov) created the Computer Crime and Intellectual Property Section (CCIPS) to deal specifically with Internet crime. This special CCIPS Website has very useful information on the following:

- Cracking policy, cyber-crime and Internet legal cases, laws, and relevant documents.
- Intellectual Property Rights (IPR) policy, cyber-crime and Internet legal cases, laws, relevant documents, and economic espionage.
- Cyber-crime and Internet documents, press releases, reports, manuals, and speeches.
- Cyber-ethics information for children, parents, and teachers as well as related Websites.

U.S. Intelligence Community

The U.S. Intelligence Community is composed of fifteen agencies, including the Central Intelligence Agency (CIA)—with its clandestine spies and numerous analysts, as well as the large Defense Intelligence Agency—specializing in intercepting global communications. Other intelligence-gathering groups in the United States include the FBI, the National Security Agency (NSA), the military (Army, Navy, Air Force, Marine Corps), the Department of State Bureau of Intelligence and Research, the Department of Energy Atomic Energy Commission, the Department of Treasury Office of Intelligence Support, and the National Imagery and Mapping Agency. Each of these agencies has its own locator information (address, telephone, fax, and Website).

Collectively, these organizations spend more than $40 billion annually. Like the CIA, they have been criticized by U.S. citizens for failing to anticipate and thwart the terrorist attacks of September 11, 2001. Also, they have been accused of incorrectly assessing the existence of banned weapons of mass destruction under former dictator Saddam Hussein's regime in Iraq, the argument posited by the U.S. and British governments for waging war in Iraq.

It seems that in recent times, the U.S. Intelligence Community has been focusing on the security of the Internet. For example, on August 15, 2005, the second-in-command public official at the Pentagon sent a letter to department leaders advising them to "Fight the Net." He said in his letter that he wants all staffers using a computer to take personal responsibility for protecting the Global Information Grid, a network connecting the Department of Defense and war-fighting systems.

Tips in the letter included using information assurance "best practices," eliminating unsecure software (like P2P file-sharing) and remote access, and minimizing access privileges with need-to-know criteria.

Independent Organizations

African Internet Community (AfriNIC)
03B3, 3rd Floor, Ebène Cyber Tower
Cyber City, Ebène, Mauritius
Tel: (230) 466-6616
Fax: (230) 466-6758
Web: www.afrinic.net

AfriNIC is a nongovernmental and not-for-profit membership-based organization. Its main role is to serve the African region as the Regional Internet Registry. The four other continents have their own Regional Internet Registries—ARIN, APNIC, RIPE, and LACNIC. Membership is open to anybody who cares to join.

American Electronics Association (AEA)
5201 Great America Parkway #520
Santa Clara, CA 95054
Tel: (408) 987–4200 Toll Free Tel: (800) 284–4232
Fax: (408) 987–4298

601 Pennsylvania Ave. NW, Suite 600, North Building
Washington, DC 20004
Tel: (202) 682–9110
Fax: (202) 682–9111
Web: www.aeanet.org/AboutAeA/AeAContactInfo.asp

The AEA was created in 1943 to be a nationwide, nonprofit trade association representing the full range of technology industry segments. It is dedicated to helping members succeed in their businesses. The association partners with small-, medium-, and large-member companies to lobby government at the state, federal, and international levels. The association also provides members with access to capital and business opportunities, as well as offering members select business services and networking programs.

American Registry for Internet Numbers (ARIN)
3635 Concorde Parkway, Suite 200
Chantilly, VA 20151–1130
Tel: (703) 227–9840
Fax: (703) 227–0671
Web: www.arin.net

The American Registry for Internet Numbers, ARIN, is a nonprofit organization. Originally it was established to administer and register Internet Protocol (IP) numbers for North America, parts of the Caribbean, and subequatorial Africa. ARIN is one of the five Regional Internet Registries collectively providing IP registration services globally—with the newest taking over the Caribbean and Africa. ARIN, it should be noted, is not an Internet Service Provider (ISP). The mission statement of ARIN notes that this organization allocates Internet Protocol resources, develops consensus-based policies, and facilitates the healthy advancement of the Internet through positive information and education.

Amnesty International
322 8th Ave.
New York, NY 10001
Tel: (212) 807–8400
Web: www.amnestyusa.org/

Amnesty International is an organization dedicated to promoting human rights in all countries. The organization is also interested in preventing abuse, particularly sexual assault and violence against women and children.

Asia Pacific Network Information Centre (APNIC)
Level 1, 33 Park Road
Milton, Brisbane, Australia
Tel: +61–7–3858–3100

Fax: +61–7–3858–3199
Web: www.apnic.net

APNIC, or Asia Pacific Network Information Centre, one of five Regional Internet Registries operating globally to register and administer IP addresses, serves the Asia Pacific region. A not-for-profit organization, its constituents consist of sixty-two economies and include Internet Service Providers and National Internet Registries. Membership in APNIC gives organizations access to all services, including requests for allocation and registration of IP address resources, as well as registration at specialized training courses. Membership also gives organizations an opportunity to participate in policy development processes and to have voting rights at membership meetings.

Center for Democracy and Technology
1634 Eye St. NW, #1100
Washington, DC, 20006
Tel: (202) 637–9800
Web: www.cdt.org/about/contact.php

The Center for Democracy and Technology promotes democratic values and constitutional liberties in the digital age. With experts in law, technology, and policy, the center searches for real-world answers to promote free expression and privacy in global communications technologies. The center brings together parties wanting to improve the future of the Internet.

CERT Coordination Center Software Engineering Institute (CERT/CC)
Carnegie Mellon University
5000 Forbes Ave.
Pittsburgh, PA 15213–3890
Tel: (412) 268–7090 (24-hr hotline)
Fax: (412) 268–6989
Web to report incidents:
    www.cert.org/reporting/incident_form.txt
Web to report system vulnerabilities:
    www.cert.org/reporting/vulnerability_form.txt

The CERT Coordination Center (CERT/CC), a center for Internet security, was created in 1988, in large part as a response to the Morris Internet worm incident. The damage incurred was the shutdown of about 10 percent of the existing Internet systems in November of that year. At the time of the incident, the Defense Advanced Research Projects Agency (DARPA) charged the Software Engineering Institute (SEI) at Carnegie Mellon University with creating a center to coordinate communication among experts, particularly during times of emergency.

Because of the rapid development of the Internet over the last decade and its growth in applications, the amount of damage that could be done and the difficulties in detecting intrusions have grown exponentially. Therefore, the role of the CERT/CC has been expanded in recent years. It has become part of the SEI Networked Systems Survivability Program—having as its main goal ensuring that appropriate technology and systems management practices are used to prevent crack attacks on networked systems. If intrusions do occur, the objective is to limit the damage incurred and to ensure continuity of critical services.

Computer and Communications Industry Association (CCIA)
666 11th St. NW
Washington, DC 20001
Tel: (202) 783–0070
Fax: (202) 783–0534
Web: www.ccianet.org/index.php

CCIA, a nonprofit organization for a wide range of companies in the computer, Internet, information technology, and telecommunications industries, was formed thirty years ago to promote open markets, open systems, open networks, and fair and open competition. The organization's goal is to protect and promote the broader interests of industry members. Today the organization assesses and shapes legislative and regulatory activities with executives from member companies—which vary in size and operate in the United States and globally. Members include computer and communications companies, equipment manufacturers, software developers, and financial service companies, as well as Internet Service Providers.

Computer Professionals for Social Responsibility (CPSR)
P.O. Box 717
Palo Alto, CA 94302
Tel: (650) 322–3778

Fax: (650) 322–4748
Web: www.cpsr.org/about

CPSR, a global organization promoting the responsible use of computer technology, was created in 1981 to educate the public and policy-makers on a wide range of topics relating to computers, professionals, and social responsibility. Projects started by CPSR include EPIC (the Electronic Privacy Information Center) and the Civil Society Project. CPSR also created the CFP (Computers, Freedom and Privacy) Conference. Although the organization was initially formed by U.S. computer scientists, it now boasts a healthy membership in more than thirty countries on six continents.

CyberAngels
P.O. Box 3171
Allentown, PA 18106
Tel: (610) 377–2966
Fax: (610) 377–3381
Web: www.cyberangels.org

In 1995 (less than two years after the World Wide Web was activated), the anticriminal activist arm of the hacker community known as the CyberAngels started to appear online. CyberAngels began after a telephone call was made to Curtis Sliwa, the founder and president of Guardian Angels and a radio talk show host for WABC in New York. At that time, a woman asked him on his talk show what he was going to do about safety in cyberspace. Sliwa faced up to the challenge by forming the first cyber-stalking help program on Internet Relay Chat.

Today the CyberAngels have more than 6,000 volunteers residing in seventy countries. Their role is to patrol the World Wide Web around the clock in the battle against child pornography and cyber-stalking. In 1998, President Bill Clinton honored the CyberAngels with the prestigious President's Service Award. CyberAngels is, in fact, the world's oldest and largest online safety organization. According to their Website, their mission is to function as a "virtual 411" safety destination. Also, the organization wants to address the Internet concerns of parents, the safety needs of children when they go online, and the prevention of online abuses and cyber-crimes—without interfering with online users' right to free speech.

Electronic Privacy Information Center (EPIC)
1718 Connecticut Ave. NW, Suite 200
Washington, DC 20009
Tel: (202) 483–1140
Web: www.epic.org/epic/about.html

EPIC, a public interest research center in Washington, D.C., was created in 1994 to focus public attention on emerging civil liberties issues and to safeguard privacy, the First Amendment, and U.S. constitutional values. The center publishes e-mail and on-line newsletters on civil liberties in the information age, reports and books on privacy, and other important topics related to civil liberties.

Free Software Foundation (FSF)
59 Temple Place, Suite 330
Boston, MA 02111–1307
Tel: (617) 542–5942
Fax: (617) 542–2652
Web: www.fsf.org

Richard Stallman, an elite hacker who was at the Artificial Intelligence (AI) Lab at MIT in the early 1970s, was the founder of the Free Software Foundation (FSF) in 1985. The FSF promotes the concept of free software—which pertains to the users' freedom to change and improve, copy, distribute, run, or study the software. Specifically, "free" applies to four types of freedom for the users of the software: (1) the ability to run the program for any function; (2) the ability to investigate how the software works and adapt it to one's own needs—with access to the source code being a precondition; (3) the ability to give copies to other users; and (4) the ability to improve the software and to release improvements to the community so that the community can benefit. This opening quotation on the FSF Website summarizes their philosophy: "Free software is a matter of liberty, not price. You should think of 'free' as in 'free speech.'"

Although other organizations freely distribute what software happens to be available, the FSF focuses on the development of new free software. The FSF also makes free software into a coherent system, thus reducing users' needs to buy proprietary software (such as that produced by the Microsoft Corporation). The FSF promotes the development and use of the GNU operating system, used widely in its GNU/Linux variant. Although the

FSF distributes copies of the GNU software and manuals for a small distribution fee, it accepts tax-deductible gifts from users to help defray the cost of further GNU development. In fact, most of FSF's operating funds are derived from distribution service revenues and gifts.

Human Rights Watch
350 Fifth Ave., 34th Floor
New York, NY 10118–3299
Tel: (212) 290–4700
Web: www.hrw.org/

Human Rights Watch, an independent organization that is non-governmental in nature, investigates and produces information on human rights abuses around the globe. The Website covers painful stories about the abuse that street children around the world suffer, including victims of cyber child pornography.

Information Technology Association of America (ITAA)
1401 Wilson Blvd., Suite 100
Arlington, VA 22209
Tel: (703) 522–5055
Fax: (703) 525–2279
Web: www.itaa.org/about/staff.cfm

ITAA is an Information Technology Trade Association that offers a wide assortment of public policy, business development, and P2P networking programs. Privacy, trust, and security issues dealing with Information Technology are of particular interest to members.

Internet Corporation for Assigned Names and Numbers (ICANN)
4676 Admiralty Way, Suite 330
Marina del Rey, CA 90292–6601
Tel: (310) 823–9358
Fax: (310) 823–8649
Web: www.icann.org

In 1998, a group of Internet stakeholders—including academics, businesses and technological experts, and online users—started the Internet Corporation for Assigned Names (ICAN) and Numbers (ICANN) to act as a technical coordination arm for the Internet. ICANN is a nonprofit, private-sector corporation. Previously, the U.S. Internet Assigned Numbers Authority (IANA) assigned the identifiers that must be globally unique for the World Wide Web to function properly. Today, ICANN manages domain names, IP addresses, protocol parameters, and port numbers. ICANN also takes care of the Internet's root server system.

Internet Engineering Task Force (IETF)
IETF Secretariat c/o Corporation for National Research
    Initiatives
1895 Preston White Dr., Suite 100
Reston, VA 20191–5434
Tel: (703) 620–8990
Fax: (703) 620–9071
Web: www.ietf.org

The IETF, or Internet Engineering Task Force, can best be described as an international community of individuals concerned with the ongoing evolution of the Internet. Network designers, vendors, operators, and academics are members, bound by an interest in the World Wide Web's architecture and smooth operation. Membership to the IETF is open, with working groups organized along the lines of routing, transport, and security.

Most member interactions are conducted by mailing lists, as meetings are held only three times annually. The working groups are managed by area directors, who are members of the Internet Engineering Steering Group (IESG). Moreover, an Internet Architecture Board (IAB) provides architectural oversight and adjudicates appeals if an individual or group complains that the IESG has not met its objectives. Both the IAB and the IESG are chartered by ISOC. The general area director acts as chair of the IESG and of the IETF and is an ex officio member of the IAB.

Internet Public Library (IPL)
University of Michigan School of Information
304 West Hall
Ann Arbor, MI 48109–1092
Web: www.ipl.org/div/contact/

The Internet Public Library (IPL) is a public service organization and learning/teaching environment located at the University of Michigan's School of Information. The IPL does the following:

- Provides library services to Internet users;
- Trains information professionals and students so they can work better in a digital environment;
- Develops technology and best practices for providing library services through the Internet, including digital reference services and collection management, and conducts research aimed at increasing knowledge about digital libraries and librarianship.

Internet Society (ISOC)
Internet Society International Secretariat
1775 Wiehle Ave., Suite 102
Reston, VA 20190
Tel: (703) 326–9880
Fax: (703) 326–9881
Web: www.isoc.org

The Internet Society (ISOC) is a professional membership society with more than 150 organizations and 20,000 individual members in over 180 nations. Its role is to provide leadership by addressing issues related to the future of the Internet. ISOC is the home for groups responsible for Internet infrastructure standards—such as the Internet Engineering Task Force (IETF). Since 1992, ISOC has served as the international organization for global coordination and cooperation on the Internet, with its primary function being the promotion of a wide range of activities related to the Internet's development, its ongoing availability, and its ever-evolving technologies.

Latin America and Caribbean Internet Addresses Registry (LACNIC)
Potosi 1517
Montevideo, 11500 Uruguay
Tel: +598 2 604222
Fax: +598 2 6042222 int. 112
Web: http://lacnic.net/sp/

The Latin America and Caribbean Internet Addresses Registry (LACNIC) is an independent, not-for-profit organization supporting the Internet's infrastructure through technical coordination in Latin America and the Caribbean. It is one of five Regional Internet Registries that provides global Internet resources and related services to members in the LACNIC service region.

As in other areas, membership consists mostly of Internet Service Providers, telecommunication organizations, and large corporations.

National Center for Missing and Exploited Children
Charles B. Wang International Children's Building
699 Prince St.
Alexandria, VA 22314–3175
Tel: (703) 274–3900
Hotline: 1–800-THE-LOST
Web: www.missingkids.com/

The National Center for Missing and Exploited Children is a private, nonprofit organization. The Website of this center provides recent and updated information on missing and abused children. The Website has many useful features, including a cyber tipline, links and realistic information for parents about keeping their children safe online, and up-to-date international coverage of child pornography. It also features timely news. For example, the Website reports that a new study of child pornography laws in 184 Interpol member countries around the world has produced alarming results: more than half of the member countries—at least ninety-five—have no laws addressing child pornography, and in many other countries, existing laws are inadequate.

National High-Tech Crime Unit (NHTCU)
PO Box 10101
London E14 9NF
Tel: +44 (0) 870 241 0549
Fax: +44 (0) 870 241 5729
Web: www.nhtcu.org/nqcontent.cfm?a_id=12261

The National High-Tech Crime Unit (NHTCU), located in the United Kingdom, has this mission posted on their Website: To combat national and transnational Internet crime, especially that which adversely impacts on the United Kingdom.

The NHTCU conducted a survey among businesses in 2003 to determine how much money they lost from Internet and computer security breaches over the previous twelve months. The NHTCU found that security breaches cost UK businesses an estimated £143 million during that period. The 105 businesses surveyed said that there were 3,000 incidents among them. The breaches included information theft, virus attacks, and the physical loss of hardware (such as laptops). Similar surveys have been jointly conducted in the United States by the FBI. As is the case with these annual U.S. surveys, a number of companies chose not to participate in the survey.

Northern Virginia Technology Council (NVTC)
2214 Rock Hill Road, Suite 300
Herndon, VA 20170
Tel: (703) 904–7878
Fax: (703) 904–8008
Web: www.nvtc.org/about/bdstaff.htm#staff

The Northern Virginia Technology Council (NVTC) is the trade association for the technology community in northern Virginia, the largest technology council in the United States. The council has more than 1,100 member companies, representing about 160,000 employees. Members include companies from all sectors of the technology industry—IT, software, hardware, the Internet, Internet Service Providers, telecommunications, bioinformatics, aerospace, and nanotechnology. Also, members include universities, embassies, nonprofit companies, and government agencies.

The organization has progressed from a small association to a mature and respected one, with partnership relations regionally, as well as at the state, national, and international levels. Its mission is to provide leadership to the technology community through networking and educational events, public policy advocacy, and promoting IT initiatives in targeted business sectors.

Privacy International (PI)
6–8 Amwell St., Clerkenwell
London, England EC1R 1UQ
Tel: +44 7947 778247
Web: www.privacyinternational.org/

Privacy International (PI), a human rights group created in 1990 as a watchdog on surveillance and privacy invasions by governments and corporations, is located in London, England, and has a U.S. office in Washington, D.C. The organization has raised awareness with campaigns and research throughout the world not only on issues related to wiretapping, national security, identification cards, medical privacy, and privacy of health information but also on freedom of information and expression in a broader sense.

Rape, Abuse, and Incest National Network (RAINN)
635-B Pennsylvania Ave., SE
Washington, DC 20003
Tel: (202) 922–1560
Web: www.rainn.org/

RAINN administers the National Sexual Assault Hotline and provides education about sexual assault to women and children. RAINN also gives information on incest prevention, prosecution, and recovery for victims. The Website has a wealth of information regarding relevant news, statistics, and links to sites discussing rape, abuse, and sexual violence. The site also has promoted special events, such as the "Race to Stop the Silence: Stop Child Sexual Abuse" on April 15, 2006, in downtown Washington, D.C.

Réseaux IP Européens (RIPE NCC)
Singel 258 1016 AB
Amsterdam, The Netherlands
Tel: +31 20 535 4444
Fax: +31 20 535 4445
Web: www.ripe.net

The RIPE NCC is an independent, not-for-profit organization supporting the Internet's infrastructure through technical coordination in Europe, the Middle East, and parts of Central Asia. It is one of five Regional Internet Registries that provides global Internet resources and related services to members in the RIPE NCC service region. As in other regions, membership consists mostly of Internet Service Providers, telecommunication organizations, and large corporations.

SANS Institute
8120 Woodmont Ave., Suite 205
Bethesda, MD 20814
Tel: (301) 654–7267
Fax: (301) 951–0140
Web: www.sans.org/

The SANS Institute is likely the largest information security training and certification source in the world. The SANS Institute develops, maintains, and makes available for free an impressive collection of research documents about information security. The SANS Institute also operates the Internet's early-warning system known as the Internet Storm Center.

The SANS (SysAdmin, Audit, Network, Security) Institute was started in 1989 as a research and education organization. Today its programs get to more than 165,000 auditors, chief information officers (CIOs), network administrators, and security professionals who share with each other lessons they have learned about information security and the Internet. Importantly, the SANS Institute tries to find solutions to the cyber challenges unveiled.

The SANS Institute shared resources include a weekly vulnerability digest (@RISK), a weekly NewsBites news digest, the Internet Storm Center warning system for the Internet, flash security alerts, and more than 1,200 award-winning research papers. During the first week of May 2005, for example, the SANS Institute warned that in the first quarter of 2005, over 600 new system vulnerabilities were detected, including flaws in products by Microsoft, Computer Associates, Oracle, McAfee and F-Secure, TrendMicro, Symantec Corporation, and some relatively new players such as RealPlayer, iTunes, and WinAmp.

Symantec Corporation
20330 Stevens Creek Blvd.
Cupertino, CA 95014
Tel: (408) 517–8000
Web: www.symantec.com

Founded in 1982, Symantec Corporation, with its headquarters in Cupertino, California, is a trusted security partner. The company has over 5,500 employees and operations in more than thirty-five countries. Considered by many to be a global leader in information security, the Symantec Corporation provides a broad range of IT security appliances, software, and services for home computer users and businesses of all sizes. Moreover, Symantec's Norton product brand is a leader in consumer security solutions.

On March 21, 2005, Symantec Corporation issued a report noting that Internet attacks grew by 28 percent in the second half of 2004, relative to the first half of the year. On average, businesses and other agencies received 13.6 attacks on their computer networks daily in the second half of 2004, they said—relative to 10.6 attacks in the first six months of that year. The financial sector was apparently the favored hack attack target. Moreover, crackers now seem to be setting their sights on mobile computers, noted the security experts at Symantec. According to this same Symantec Corporation report, the favored attack tools included adware and spyware, as well as phishing. The reported cost to U.S. firms in 2004 from phishing scams alone was in excess of $1.2 billion.

U.S. Council of International Business (USCIB)
1212 Avenue of the Americas
New York, NY 10036
Tel: (212) 354–4480
Policy Advocacy Fax: (212) 575–0327
Membership Fax: (212) 391–6568
Web: www.uscib.org/

The U.S. Council for International Business (USCIB) presents American business ideas, values, and answers to problems on wide-ranging issues, including the environment, e-commerce, and labor relations. Relationships exist with U.S. policy-makers and officials in the United Nations, the European Union (the EU), and a number of other governments and groups. Originally founded in 1945 to promote free trade and to help represent business in the United Nations, USCIB is currently viewed as a global network of industry partners having a fine reputation for giving reliable policy advice. The group is particularly interested in exchanging views on regulatory issues and best business practices globally.

U.S. Internet Caucus
Eye St. NW, #1107
Washington, DC 20006
Tel: (202) 638–4370
Fax: (202) 637–0968
Web: www.netcaucus.org/contact.shtml

This caucus is a broad and diverse group of public interest, nonprofit, and industry groups that wish to educate the Congress and the public about important Internet-related policy issues.

U.S. Internet Service Provider Association (USISPA)
1330 Connecticut Ave. NW
Washington, DC 20036

Tel: (202) 862–3816
Fax: (202) 261–0604
Web: www.cix.org/contactusispa.html

The U.S. Internet Provider Association (USISPA) acts as the Internet Service Provider's community representative during policy debates as well as a venue where members can exchange information and develop best practices for handling Internet-related legal issues. Although the association's principal focus is on legal and policy matters in the United States, the organization also monitors and engages in international developments. Also, the organization focuses on U.S. legal and federal matters; when necessary, it addresses state-level issues.

World Wide Web Consortium (W3C)
Massachusetts Institute of Technology (MIT)
Computer Science and Artificial Intelligence Laboratory (CSAIL)
200 Technology Square
Cambridge, MA 02139
Tel: (617) 253–2613
Fax: (617) 258–5999
Web: www.w3.org

Tim Berners-Lee started the World Wide Web (W3C) Consortium at the MIT Laboratory for Computer Science in 1994. He collaborated with CERN, a European Organization for Nuclear Research, the world's largest particle physics center and the creative inventor of many advances in information technologies. The function of the W3C is to promote interoperability and to provide a forum for discussing the World Wide Web's ongoing evolution and development. To that end, W3C promotes and develops its vision for the future of the Web, designs technologies to realize this vision, and standardizes Web technologies (which is what interoperability is all about).

Academic Degrees in Internet Security

A number of universities in the United States and elsewhere have identified Internet security as an independent area of study and specialization. Thus there are a growing number of dedicated undergraduate and graduate degrees with an Information Technology and Internet Security specialization. Most of these require the student to have residency on campus. Some of these educational institutions are listed below. The Liverpool University, however, offers a fully online master's degree on the Internet.

Idaho State University
Computer Information Systems Department
921 S. 8th Ave.
Pocatello, ID 83209
Tel: (208) 282–3585
Web: www.isu.edu/

James Madison University
Commonwealth Information Security Center
800 S. Main St.
Harrisonburg, VA 22807
Tel: (540) 568–6211
Web: www.jmu.edu/

Mary Washington College
James Monroe Center
1301 College Ave.
Fredericksburg, VA 22401
Tel: (540) 654–1000
Web: www.collegeprofiles.com/marywashington.html

University of Advancing Technology (UAT)
2625 W. Baseline Rd.
Tempe, AZ 85283–1042
Tel: (602) 383–8228
Web: http://www.uat.edu/

University of Liverpool
Liverpool, England L69 3BX
Tel: +44 (0) 151 794 2000
Web: www.liv.ac.uk/contacts/index.htm

The University of Liverpool is internationally recognized for excellence in teaching, research, and innovation. It is a member of the elite Russell Group of nineteen research-driven UK universities, including Oxford and Cambridge. It was awarded a Royal Charter in 1903. The university offers a totally online master of science degree in IT-Internet Computing. Study modules include

Internet and Multimedia Technology, Web XML Applications, various programming languages, and applications.

University of Ontario Institute of Technology (UOIT)
Faculty of Business and Information Technology
2000 Simcoe St. N.
Oshawa, ON, Canada L1H 7K4
Tel: (905) 721–8668
Web: www.uoit.ca

This Canadian university offers a bachelor of information technology degree in networking, security, and game development entrepreneurship, as well as a master's in IT security. At the present time, some residency is required.

8

Selected Print and Nonprint Resources

Resource materials about Internet crime have focused over the past several years on such topics as identity theft, cracking, spamming, and Intellectual Property (IP) infringement. Because of citizens' and industries' concerns about such cyber-threats, resource materials related to these topics have become increasingly abundant. This chapter focuses on three types of resource materials available to the public: books detailing social and technical aspects of such Internet crimes, company Websites offering protection for such problems, and films capitalizing on cyber-threat story lines.

Because every day new Internet crimes are reported by the U.S. Department of Justice's Intellectual Property Section, individuals wanting to keep abreast of new cases and laws can do so by visiting the U.S. Department of Justice's Website <www .usdoj.gov/criminal/cyber-crime/index.html>. Besides offering resource materials on cracking, Intellectual Property infringement, Internet crime documents, and cyber-ethics topics, this government Website allows individuals wanting specialized Internet abuse information to receive it through their e-mail. The following key words can be entered for specific information searches of interest to these online user segments:

- Parents, teachers, and students
- Police officers and law enforcement agents
- Citizens wanting to know more about Internet and cyber policies
- Targets of Internet crime
- High-tech industry experts

- Media professionals
- Citizens wanting to know more about Internet and high-tech privacy issues

Finally, this U.S. Department of Justice Website has some very practical Internet crime information, such as how to report Internet-related crimes; how industry partners can help fight Internet abuses; law enforcement coordination of Internet-related crimes; legal issues dealing with e-commerce, encryption, and Internet crime; federal and state laws related to Internet crime; Intellectual Property laws and breaches of those laws; international aspects of Internet crime (such as the Council of Europe Convention on Cyber-crime); how to protect critical infrastructures; the free speech issues of controversy; and privacy, security, and trust issues related to the Internet.

Books

Anguilla, J., and D. Ronfeldt. 2001. *Networks and Netwars: The Future of Terror, Crime, and Militancy.* Santa Monica, CA: Rand.
This book presents a fascinating new spectrum of cyber-conflict. Among other topics, the books discusses netwars—conflicts that terrorists, criminals, gangs, and ethnic extremists wage—and how to combat them.

Bakardjieva, M. 2005. *Internet Society: The Internet in Everyday Life.* Thousand Oaks, CA: SAGE.
Bakardjieva examines how people integrate the Internet into their ordinary lives and adapt their activities, as appropriate, to cope with technology.

Benedict, H. *Recovery.* 1994. New York: Columbia University Press.
This book is a sound resource for information on surviving and healing from rape and sexual assault. Very useful information is given for victims, family members, and friends.

Berkowitz, B. 2003. *The New Face of War: How War Will Be Fought in the 21st Century.* New York: Simon and Schuster.
With the growth of the Internet in developing nations, Berkowitz presents a fascinating treatment of how information

wars have revolutionized modern-day combat. The book addresses the important topic of how the war against cyber-terrorists can be fought—and won.

Blane, J. 2003. *Cybercrime and Cyberterrorism: Current Issues.* Comack, NY: Nova Science.

With the evolution of the Internet, new ways of committing old-fashioned crimes like fraud, stalking, and identity theft have emerged. This book discusses some very important topics on cyber-crime and cyber-terrorism—including how those two concepts are quite distinct.

Brill, A., F. Baldwin, and R. Munro. 1998. *Cybercrime and Security.* New York: Oceana.

This three-binder set presents a rather advanced publication on cyber-crime and security. Prepared by leading experts in these fields, the book gives useful management strategies and solutions for system administrators who want to keep their networks safe. The book also alerts readers about potential threats to networks, discusses pertinent cyber-crime laws, and details privacy issues, as well as encryption and computer security.

Casey, E. 2000. *Digital Evidence and Computer Crime.* San Diego, CA: Academic.

Meant primarily for a legal audience, this book discusses legal aspects of computer network crimes. It describes how evidence stored on or transmitted by computers can play a role in the ever-growing field of cyber-crime, including stalking, online harassment, fraud, theft, terrorism, and child pornography.

Castells, M. 2003. *Internet Galaxy: Reflections on the Internet, Business, and Society.* New York: Oxford University Press USA.

This book speculates about how the Internet will change aspects of human life—such as politics, businesses, and economies in the future. This is a nontechnical book intended for professionals and graduate/undergraduate students of sociology and computer science. One of the major contributions of this book is its ability to point out the Internet's capacity to liberate. It also points out that the Internet can marginalize those who have no access to it. Another important factor addressed in this fascinating work is the Internet's effect on interpersonal relationships.

Berkowitz, B. 2000. *Rise of the Network Society.* [First volume of trilogy: *The Information Age: Economy, Society, and Culture*]. Malden, MA: Blackwell.

This book presents an analysis of the economical, political, and social effects of Information Technology (IT) in our contemporary world. One of the main points of this work is the potential exclusion of information to countries with limited access to the Internet, while in developed countries, the flow of information exchange in the new age of information increases at unprecedented rates. Castell's intended audience is professionals and college/university students of sociology and computer science.

Bragg, R., M. Rhodes-Ousley, K. Strassberg, and B. Buege. 2004. Network Security: *The Complete Reference*. New York: McGraw Hill-Osborne.

A complete review of network-security issues that will be helpful to amateurs and experts alike. Especially good for those interested in risk management and the creation of policies related to network security.

Chirillo, J. 2001. *Hack Attacks Encyclopedia: A Complete History of Hacks, Phreaks, and Spies over Time.* New York: John Wiley and Sons.

This book, written by a computer security expert, contains historic texts, program files, code snippets, cracking and security tools, and some quite advanced topics—such as password programs, Unix and Linux systems, scanners, sniffers, spoofers, and flooders. This book is not intended for a neophyte audience.

Clifford, R. 2001. *Cybercrime: The Investigation, Prosecution, and Defense of a Computer-Related Crime.* Durham, NC: Carolina Academic.

This book was written primarily for a legal audience. To that end it presents legal topics such as what conduct is considered to be a cyber-crime, investigating improper cyber-conduct, trying a cyber-crime case as a prosecuting or defending attorney, and handling integral international aspects of cyber-crimes, particularly when more than one jurisdiction is involved.

Cole, E., and J. Riley. 2001. *Hackers Beware: The Ultimate Guide to Network Security.* Upper Saddle River, NJ: Pearson Education.

This book, penned by experts in computer security, is intended for the advanced network security professional. To that end, it describes UNIX and Microsoft software vulnerabilities. It also discusses protection against system intrusions. Also, it documents trends and gives critical thoughts regarding system administration, network, security, and trust.

Feinstein, D. 2003. *Improving Our Ability to Fight Cybercrime: Oversight of the National Infrastructure Protection Center: Congressional Hearing.* Collingdale, PA: Diane.

This current book discusses in frank terms how the United States needs to protect its citizens following the terrorist attacks on the World Trade Center and the Pentagon on September 11, 2001. Important issues center on the National Infrastructure Protection Center.

Furnell, S. 2001. *Cybercrime: Vandalizing the Information Society.* Reading, MA: Addison-Wesley.

This book is useful because one does not have to be a cyber expert to understand what the author is relating. Written by a British computer security expert, it gives a thorough overview of cracking, viral code, and electronic fraud. It details a wide range of cyber-crimes and abuses.

Garfinkel, W., G. Spafford, and D. Russell. 2001. *Web Security, Privacy, and Commerce.* Sebastopol, CA: O'Reilly and Associates.

Given the importance of electronic business in society today, this book does a fine job of detailing issues related to Web-based security, privacy, and online commerce. Topics include public key infrastructure, digital signatures, digital certificates, hostile mobile code, and Web publishing.

Goodman, D., and A. Sofaer. 2001. *The Transnational Dimension of Cybercrime and Terrorism.* Prague: Hoover Institute.

Given the importance of transnational issues in cyberspace, this book is written to help readers better understand its complexity. Interesting portions cover cyber-terrorism.

Gunkel, D. 2000. *Hacking Cyberspace.* Boulder, CO: Westview.

This book is intended for a rather advanced audience, since it examines in some detail the metaphors of new technology and how those metaphors affect the implementation of technology in today's World Wide Web. The book touches on philosophy, communication theory, and computer history to tell its important story.

Himanen, P., M. Castells, and L. Torvald. 2001. *The Hacker Ethic and the Spirit of the Information Age.* New York: Random House.

This book is one of the few that offers insights into White Hat hackers and the positive hacker ethic that motivates them to do what they are, technically speaking, very capable of doing. It talks about how White Hats from the 1960s through the present have been able to create great things by joining forces and using information in very creative ways.

Juergensmeyer, M. 2000. *Terror in the Mind of God.* Berkeley: University of California Press.

Focusing on recent events not unlike the Muslim cartoon that caused havoc in parts of the world in 2006, the first part of this book explores the use of violence by marginal groups within five major religions. The author details the theological justifications for violence and the bases for its authorization. The second part of this book describes common themes and patterns in the cultures of violence detailed in the first part. This intriguing book closes with suggestions for the future of religious violence.

Levy, S. 2001. *Hackers: Heroes of the Computer Revolution.* New York: Penguin.

This book is meant to be a positive influence on those interested in hacking for all the right reasons; it describes, in a very user-friendly way, the pranks played at MIT's Tech Model Railroad Club and the very famous computer security experts who had their roots planted there. This is a must-read for those wanting to know about how the Computer Underground was started.

Lilley, P. 2003. *Hacked, Attacked, and Abused: Digital Crime Exposed.* London: Kogan Page.

Given that the business audience is concerned about keeping networks safe from intrusions, this book provides sound, practical advice in that regard. Topics include digital crime, cyber-laundering, fraudulent Internet Websites, viruses, Website defacement, identity theft, information warfare, Denial of Service (DoS) attacks, and digital privacy invasion.

Loader, B., and T. Douglas. 2000. *Cybercrime: Security and Surveillance in the Information Age.* New York: Routledge.

This pair of authors focus on the increasing concern experts have over the use of electronic communications for criminal activities and the controversies surrounding countermeasures currently used by governments to combat cyber-crime. This book covers a wide range of topics, including legal, psychological, and sociological aspects of cyber-crime.

Maiwald, E. 2001. *Network Security: A Beginner's Guide.* New York: McGraw-Hill.

Despite its soothing title, this book is quite far from a beginner's textbook on network security. The book seems to be written for network administrators running a network but needing to secure it as well. Topics include antivirus software, firewalls, and intrusion detection systems (IDS).

McClure, S., S. Shah, and S. Shah. 2002. *Web Hacking: Attacks and Defense.* Upper Saddle River, NJ: Pearson.

This interesting book discusses what happens when vulnerabilities in networks go unrepaired. It is an informative guide for Web security professionals.

McIntosh, N. 2002. *Cybercrime.* Chicago: Heinemann.

This is one of a few books on the market that gives rational but rather elementary information to those wanting to know more about cyber-crime. It is written so that those students aged nine through twelve will understand its main points.

Meinel, C. 2001. *The Happy Hacker.* 4th ed. Show Low, AZ: American Eagle.

This book instructs readers on how to hack their computers—safely. Especially useful for newcomers to the hacking field, this book teaches in simple language how to get on the Internet like an experienced hacker, how to map the Internet, how to find computers to crack, how to construct e-mail bombs, and how to track Black Hats and spammers on the Internet. Other unique topics include "how to meet other hackers" and "hacker humour."

Mitnick, K., and W. Simon. 2002. *The Art of Deception: Controlling the Human Element of Security.* New York: John Wiley and Sons.

Infamous cracker Kevin Mitnick, who spent a number of years behind prison bars for his cracking exploits, joins forces with William Simon to write about the importance of social engineering in gaining successful crack attacks. These two authors also offer some useful advice to readers about securing business computer networks. The two authors also have a similar but updated 2005 book release titled *The Art of Intrusion: The Real Stories behind the Exploits of Hackers, Intruders, and Deceivers.*

Newman, J. 1999. *Identity Theft: The Cybecrime of the Millennium.* Port Townsend, WA: Loompanics.

The title of the book describes the important topic of concern for citizens around the globe: identity theft. It gives a nontechnical treatment of the topic and deals with the complexity of the issue in the United States, in particular.

Nichols, R., and P. Lekkas. 2002. *Wireless Security: Models, Threats, and Solutions.* New York: McGraw-Hill.

Given the rise in Internet and wireless cracking exploits in recent years, these authors do a fine job of presenting some technical solutions for voice, data, and mobile commerce breaches.

Raymond, E. 1996. *The New Hacker's Dictionary.* Boston: MIT Press.

This book defines the words used by computer hackers and programmers from the start of the underground movement until the mid-1990s. It is required reading for anyone caring about the history of the computer industry and the dynamics of the information economy.

Richards, J. 1998. *Transnational Criminal Organizations, Cybercrime, and Money Laundering: A Handbook for Law Enforcement Officers, Auditors, and Financial Investigators.* Boca Raton, FL: CRC.

Given the transnational nature of cyber-crime, this book is an essential read for those in the law enforcement profession. The book covers the workings of organized criminals and groups transcending national borders. Topics include how criminals internationally launder money, how law enforcement curbs such activities, and new methods of counteracting cross-border crimes, including those using the Internet.

Schell, B., and J. Dodge, with S. Moutsatsos. 2002. *The Hacking of America: Who's Doing It, Why, and How.* Westport, CT: Quorum.

Written by two university professors and a lawyer specializing in cyber-crime, this book used psychological inventories to study more than 200 hackers and crackers who attended the Def-Con and Hackers on Planet Earth (HOPE) hacking conventions in 2000. The personalities and behavioral traits of male and female hackers are detailed, and an intriguing question—Is the vilification of hackers justified?—is answered. Very interesting case studies on hackers—young and old—are presented.

Schell, B., and N. Lanteigne. 2000. *Stalking, Harassment, and Murder in the Workplace: Guidelines for Protection and Prevention.* Westport, CT: Quorum.

This book provides a clear, objective, responsible, and readable analysis of the facts of stalking crimes, including cyber-stalking. It also provides a practical guide for individuals and organizations trying to protect themselves from the harms associated with stalking and cyber-stalking. It provides insights into why stalkers and cyber-stalkers behave the way they do.

Schell, B., and C. Martin. 2004. *Contemporary World Issues: Cybercrime.* Santa Barbara, CA: ABC-CLIO.

This book examines many forms of computer exploits—some positively motivated by the White Hats—and some negatively motivated by the Black Hats—from the Hacking Prehistory Era before 1969 through the present. Interesting topics include social engineering and its importance in cracking, famous cyber-crime exploits targeting property and persons, and the technical but understandable differences between cracking, flooding, virus and worm production and release, spoofing, phreaking, and piracy. Solutions for dealing with these exploits are also addressed.

Schell, B., and C. Martin. 2006. *Webster's New World Hacker Dictionary.* Hoboken, NJ: Wiley.

This dictionary has about 900 entries on important people, terms, laws, and references related to hacking and to Internet-related crime. A preface, timeline on critical cracking and technology improvement events, and an appendix by Carolyn Meinel on "How Do Hackers Break into Computers?" are included. Some interesting legal cases on crackers and cyber-criminals are also outlined.

Schneier, B. 2000. *Secrets and Lies: Digital Security in a Networked World.* New York: John Wiley and Sons.

Written by an information security expert and intended for a more mature business audience, this book tells industry leaders what they must know about computer security in order to keep the business prepared for cyber-intrusions and disaster recovery. Although for the most part this book is rather technical in focus, it does give some insights into the digital world. Also, the realities of a networked society—good and bad—are illustrated through interesting sidebars.

Shimomura, T., and J. Markoff. 1996. *Takedown: The Pursuit and Capture of Kevin Mitnick, America's Most Wanted Computer Outlaw, by the Man Who Did It.* New York: Warner.

This book is a fascinating read, for it describes the capture of one of America's most wanted Internet criminals, later turned IT security guru, Kevin Mitnick. It includes some details of Shimomura's personal life as a White Hat hacker, as well as some of the technical, legal, and ethical questions regarding the case.

Shinder, D., and E. Tittel. 2002. *Scene of the Cybercrime: Computer Forensics Handbook.* Rockland, MA: Syngress.

The purpose of this book is to introduce IT experts responsible for building and maintaining computer systems to stop cyber-crime by giving them an understanding of the principles of law enforcement. Also, law enforcement officers are helped to gain an understanding of the technical aspects of cyber-crime and how technology can be applied to solve such crimes.

Smith, M. 2004. *Encyclopedia of Rape.* Westport, CT: Greenwood.

This book takes a novel approach to the examination and understanding of rape and sexual assault, including that involving children. Besides discussing pertinent issues regarding sexual assault, the authors detail the places, laws, and customs globally regarding rape and sexual assault. Also, literature and artistic depictions of rape and sexual assault are described, as well as the social and political events concerning these heinous crimes. Related readings are given, along with encyclopedia entries.

Spinello, R., and H. Tavani. 2001. *Readings in CyberEthics.* Boston: Jones and Bartlett.
    This anthology of more than forty essays presents some very interesting but conflicting points of view about evolving moral and ethical questions raised by the Internet. Topics include free speech versus content controls and censorship, Intellectual Property infringement versus open-source access, and online professional ethics and codes of conduct.

Tavani, H. T. 2004. *Ethics and Technology: Ethical Issues in an Age of Information and Communications Technology.* Hoboken, NJ: Wiley.
    A good, lay-oriented resource.

Westby, J. 2003. *International Guide to Combating Cybercrime.* Chicago: ABA.
    Giving the international nature of cyber-crime, this book is timely and important. It discusses the complex jurisdictional issues pertaining to the curbing of international crime related to the Internet.

Internet-Related Journals and Magazines

Journals

Academic Open Internet Journal
http://www.acadjournal.com/Default.asp

This online journal was created by Sotir Sotiroy and Andrey Nonov in April 2000 to publish papers from scientists around the globe promoting technical advances of the century. The journal's motto states: "Let's make science closer to Information technologies." To that end, the journal promotes the use of the Internet and the development of network resources to help scientists to communicate. Topics of interest include new techniques and databases advancing the evolution of science and the Internet.

Digital Humanities Quarterly
http://www.noraproject.org/dhq/index.shtml

This new journal was first published in March 2006. This quarterly, meant to be an open-access, peer-reviewed digital journal covering the full range of digital and Internet-driven media in the humanities—including scholarly articles, experiments in interactive media, and reviews of books, Websites, and new media art installations—is published by the Alliance of Digital Humanities Organizations (http://www.digitalhumanities .org/). The alliance also supports other Internet-related journals such as *Literary and Computing Linguistic* (http://llc .oupjournals.org/), a print journal published by Oxford University Press.

EJournal
http://www.ucalgary.ca/ejournal/

This rather broad-based electronic journal is concerned with the implications of electronic networks and texts. Produced since 1991, the journal publishes articles on both theories and practices relating to the production, transmission, storage, interpretation, alteration, and replication of electronic text—including the full range of social, psychological, literary, and economic implications of computer networks and the Internet. An interesting article featured in the September 2004 issue was titled "What Makes Web Pages Different: Reemphasizing the Role of Hypertext to Develop Rhetorical Webs."

Information Research
http://informationr.net/ir/

This journal is an open-access, international scholarly journal, created to make accessible to readers around the globe the results of research regarding a full range of information-related disciplines and including topics relating to the transmission of information over the Internet. The journal is edited by Professor Tom Wilson and is hosted by Lund University Libraries, the

Swedish School of Library and Information Science, and Gothenburg University in Borås.

Information Society Journal
http://www.indiana.edu/~tisj/

This journal has been published since 1981 and is considered to be an important forum for the cutting-edge analysis of the impacts, policies, and methodologies relating to information technologies and the changes that they bring to society and culture. Topics of interest include computers and telecommunications, the Internet, and innovative social forms in cyberspace. A refereed journal, it originates in the Department of Telecommunications at Indiana University.

International Journal on Digital Libraries
http://link.springer.de/link/service/journals/00799/index.htm

This international quarterly journal was created to advance the theory and practice of the acquisition, definition, organization, and management and dissemination of information through global networking, including the World Wide Web. To that end, topics of interest include digital information production and management in high-speed networks, the seamless integration of information over networks and the Internet, the security and privacy of individuals and of business transactions conducted over the Internet, and designing efficient, effective business processes involving the Internet.

Internet and Higher Education
http://www.elsevier.com/locate/iheduc

This quarterly journal was created to address modern-day issues and future developments related to online learning, teaching, and the administration of the Internet in postsecondary settings. This peer-reviewed journal publishes articles addressing innovative deployments of Internet technology in instruction and, in particular, on research demonstrating the effects of the Internet and Information Technology on instruction in the postsecondary classroom. The journal is both international and interdisciplinary in focus, with topics including online teaching, learning, management, and administration; Internet technology design and use; and issues and trends in synchronous, asynchronous, and hybrid online learning.

The Internet Journal of Allied Health Sciences and Practice
http://ijahsp.nova.edu/

This quarterly, online, peer-reviewed scholarly journal is dedicated to the exploration of allied health professional practice and education using modern technologies, such as the Internet, for communications. The journal was established to meet a growing interest in the advancement of allied health professionals as integral members of the health care team, and to that end the journal provides an interdisciplinary forum in which allied health professionals may share initiatives involving science, practice, and education.

Journal of Internet Banking and Commerce
http://www.arraydev.com/commerce/jibc/index.htm

This journal, published since 1996, was created to inform banking and e-commerce academics and executives about key developments in electronic commerce, profit-oriented business architecture, key e-commerce benchmarking practices, risk and accountability, and present and future trends in Internet-based e-commerce and banking. This free online journal acts as a vehicle for banking professionals to discuss critical e-commerce issues through published articles and to establish research and business contact networking. To that end, the journal publishes original articles describing e-commerce theory and practice, important announcements, guest columns, letters to the editor, and surveys advancing e-commerce and banking. In the April 2006 issue, for example, the journal published an interesting article by Lior Shamir of Michigan Tech University, titled "A Security Concern in MS-Windows: Stealing User Information from Internet Browsers Using Faked Windows."

Journal of Internet Technology and Secured Transactions
http://www.inderscience.com/browse/index.php?journal
    CODE=ijitst

This journal, published four times a year, is an international journal dedicated to the advancement of theory and the practical implementation of secured Internet transactions. It strives to foster informed discussions on Information Technology and the Internet's evolution. The objective of the journal is to promote World Wide Web excellence by bridging the knowledge gap between scholars and industrial practitioners by promoting the exchange of ideas aimed at secured Internet transactions and encompassing conceptual analysis, design implementation, and performance evaluation.

Michigan Telecommunications and Technology Law Review
http://www.law.umich.edu/JournalsandOrgs/mttlr.htm

This journal is said to be one of the first law journals to capitalize on the use of interactive media to promote informed discussions and debates about the intertwined legal, social, business, and public policy issues raised by emerging technologies, such as the Internet. Moreover, the journal is concerned, in general, with the emergence of new techniques and technologies in computing, telecommunications, multimedia, networking and information services, and, in particular, with the tensions created within society as a result of such changes. Journal topics include administrative law, communications and constitutional law, Intellectual Property Rights law, and international law. In March 2006 the journal hosted a symposium titled "21st Century Copyright Law in the Digital Domain," a conference meant to bring together scholars, legal practitioners, policy-makers, and technologists to address the legal and policy questions brought about by major changes in copyright law as a result of the Internet's evolution.

Public-Access Computer Systems Review
http://info.lib.uh.edu/pr/pacsrev.html

Although no longer in operation, this electronic journal was focused on end-user computer systems in libraries. Established in 1989 by Charles Bailey, Jr., who was the editor-in-chief until 1996, the journal was published by the University of Houston Libraries three times a year. In 1991 the journal moved to a double-blind review process, publishing articles on broad network-based information issues. For example, one of the editor-in-chief's published articles was titled "Network-based Electronic Publishing of Scholarly Works: A Selective Bibliography." After Charles Bailey's reign ended, Pat Ensor and Thomas Wilson became the editors-in-chief in January 1997. The journal's closing announcement was made on June 18, 1998, but submitted papers remained under consideration for publication until August 2000, when the journal ceased operations altogether. During its nine years of publication, the journal published forty-two issues that included more than 100 articles, columns, reviews, and editorials on network-based information topics. Readers can still access these published articles.

Richmond Journal of Law and Technology
http://law.richmond.edu/jolt/index.asp

This journal was purportedly one of the first law reviews in the world to be published exclusively online. First published on April 10, 1995, the journal focused on the impact that computer-related and other emerging technologies have on law. Today, the journal's articles reach thousands of readers per month in more than seventy countries, with four or five issues being published per year. Journal topics include the ongoing legal issues related to the Internet and to issues in biotechnology and the emerging areas of constitutional and technology law.

Magazines

Byte Magazine Online
http://www.byte.com/byte.htm

This magazine, with Jonathan Erickson as the editorial director, features news items and up-to-date articles dealing with technology-related topics, including the Internet.

Computers & Texts
http://users.ox.ac.uk/~ctitext2/publish/comtxt/

This magazine, with editor Michael Fraser, was formerly published between March and August of each year in the United Kingdom and dealt with articles relating computers to texts. The journal ceased publication with volume 19.

Java Developer's Journal
http://www.sys-con/com/java/

This online magazine is devoted to pieces of interest to Java language developers and provides updates on the Java programming language and its applications.

Linux Journal
http://www.linuxjournal.com/

Since 1994, this online magazine has provided monthly news and pieces of interest to the Linux community.

Network Magazine Online
http://www.itarchitect.com/

This online magazine, now called *IT Architect*, provides news and practical articles related to standards and technologies of interest to the IT architect community.

PC WORLD.COM
http://www.pcworld.com/

This online magazine publishes topics relating to Personal Computers (PCs) and modern-day computer technology. Individuals having a magazine subscription can have full and free access to the most current online magazine edition, as well as to all of the copy archives dating back to 1998.

Perl Journal
http://www.tpj.com/

This journal, founded in 1996 by Jon Orwant, was published until January 2006. It was considered to be a leading publication for and about Perl programming. Archives of the magazine can be found at http://www.tpj.com/. Individuals wishing to submit articles on Perl or any other Lightweight Language software development are encouraged to do so in the Dr. Dobb's Portal, known as the Lightweight Languages Department and located at http://www.ddj.com/dept/lightlang/;jsessionid=WXDFOU 5BSLUHSQSNDBCCKH0CJUMEKJVN.

TechWeb
http://www.techweb.com/

This online magazine, dedicated to business network professionals, has news and up-to-date Internet articles related to electronic business transactions and online security topics such as identity theft.

Ubiquity
http://www.acm.org/ubiquity/homepage.html

This online magazine, an ACM (Association for Computing Machinery) Information Technology magazine, presents articles and news features of interest to Information Technology specialists. For example, the May 2006 featured article had a quotation by Espen Andersen of the Norwegian School of Management, who claimed that the Internet is the fastest medium of all time, swarming with updates, links, and searchability features. Influence on the Internet, he affirmed, is determined primarily by its readers, who, using dialogue and references, feed priorities to the search engines. Media companies who do not shape their product to this evolution, warned Andersen, will gradually lose their ability to decide what is important.

UnixReview.com
http://www.unixreview.com/

This online magazine publishes articles on computer server management and features updates on programming languages (such as Unix, Perl, Python, and Shell) and on operating systems (such as Linux and Solaris).

WindoWatch
http://www.windowatch.com/

This online magazine closed its digital doors in January 2006. Lois Bleichfeld Laulicht, the magazine's editor and founder, notes on the Website that the first issue was posted in October 1994 and remained in operation for eleven years. The objective of the magazine was to bring to the public well-informed discussions by those understanding computers and their systems, as well as those having broad and informed interests in less "techie" subjects.

ZD Net
http://www.zdnet.com/

This online magazine has news, blogs, and white papers dealing with secure transactions over the Internet. Moreover, the magazine's white paper directory is said to be the Web's largest library containing free Information Technology white papers and numerous case studies on data management, IT management, networking, communications, enterprise applications, and IT privacy, security, and trust issues.

Online Newspapers, Scholarly and Literary Journals, and E-Zines

The Internet provides online users with access to numerous newspapers, scholarly and literary journals, and e-zines (that is, electronic magazines) around the globe—just with the click of a computer mouse. Below is a list of URLs to such online resources, meant to be illustrative of the wide-ranging availability of online popular media. It is interesting to note that although the online versions of newspapers and scholarly and literary journals are not unlike those available in hard copy, the e-zines, in particular, tend to be geared toward the arts and fiction communities as well as hacktivists and, therefore, are often quite controversial in nature.

URLs of Selected Online Newspapers

AlterNet. http://www.alternet.org/
Arts & Letters Daily. http://www.aldaily.com/
The Atlantic. http://www.theatlantic.com/
Austin American-Statesman. http://www.austin360.com/aas/

Boston Globe. http://www.boston.com/globe/
Boston Review. http://bostonreview.net/
British Media Online. http://www.wrx.zen.co.uk/britnews.htm

Chicago Tribune. http://chicagotribune.com/
Christian Science Monitor. http://www.csmonitor.com/
Chronicle of Higher Education. http://chronicle.com/
Commentary. http://www.commentarymagazine.com/
Common Dreams. http://www.commondreams.org/
CounterPunch. http://counterpunch.com/
CNN Web. http://www.cnn.com/
Cursor. http://www.cursor.org/toc.htm

Daily Mirror. http://www.mirror.co.uk/
Daily Newspaper.co.uk. http://dailynewspaper.co.uk/
Dallas Morning News. http://www.dallasnews.com
Daypop. http://www.daypop.com/
Detroit News Online. http://www.detnews.com/

EBR: Electronic Book Review. http://www.altx.com/ebr/
Edge. http://www.edge.org/

Guardian. http://www.guardian.co.uk/

HotWired. http://www.hotwired.com/
Houston Chronicle. http://houstonchronicle.com/
Human Events Online. http://www.humaneventsonline.com/

Independent. http://news.independent.co.uk/
Information Clearing House. http://www.informationclearing-
     house.info/index.htm/

London Review of Books. http://www.lrb.co.uk/
Los Angeles Times. http://www.latimes.com/

Media Times Review. http://www.mediatimesreview.com/
     indexEN.php
Miami Herald. http://www.miami.com/mld/miamiherald/
MotherJones. http://www.motherjones.com/index.html

Nation. http://www.thenation.com/
National Review Online. http://www.nationalreview.com/
New Republic. http://www.thenewrepublic.com/
New Yorker. http://www.newyorker.com/
New York Review of Books. http://www.nybooks.com/
New York Times. http://www.nytimes.com/

Palo Alto Weekly. http://www.service.com/PAW/home.html
Philadelphia Inquirer. http://www.philly.com/mld/inquirer/

Reason Online. http://reason.com/
RollingStone. http://rollingstone.com/

Salon. http://www.salon.com/
San Antonio Express-News. http://www.express-news.com/
Schweriner Volkszeitung. http://www.svz.de/schwerin.html
Slate. http://www.slate.msn.com/
Spiegel, Der. http://www.spiegel.de/
SunSpot. http://www.sunspot.net/

Tatler. http://www.tatler.co.uk/

Texas Monthly. http://www.texasmonthly.com/
The Times. http://www.the-times.co.uk/
TLS: Times Literary Supplement. http://www.the-tls.co.uk/
Times-Picayune. http://www.timespicayune.com/

U.S. News & World Report. http://www.usnews.com/

Village Voice. http://www.villagevoice.com/

Wall Street Journal. http://public.wsj.com/home.html
Washington Post. http://www.washingtonpost.com/
Washington Times. http://www.washingtontimes.com/
WEEK. http://metrotel.co.uk/theweek/
Weekly Standard. http://www.weeklystandard.com/

URLs of Selected Scholarly and Literary Journals

American Drama. http://www.americandrama.org/
American Journalism Review. http://www.ajr.org/
American Political Science Review. http://www.jstor.org/
journals/00030554.html
American Quarterly. http://muse.jhu.edu/journals/aq/
ANQ: A Quarterly Journal of Short Articles, Notes, and Reviews. http://www.heldref.org/html/body_anq.html
Anthropoetics. http://www.humnet.ucla.edu/humnet/anthro
poetics/home.html
APS Publishing News—Online. http://publish.aps.org/
Arachnion: A Journal of Ancient Literature and History. http://
www.cisi.unito.it/arachne/arachne.html
Arid Lands Newsletter. http://ag.arizona.edu/OALS/ALN/
ALNHome.html
Arts Education Policy Review. http://heldref.org/html/body_
aepr.html
Ascending Node. http://seds.lpl.arizona.edu/nodes/ass_node
.html
Asian Affairs. http://heldref.org/html/body_aa.html
Australian Humanities Review. http://www.lib.latrobe.edu.au/
AHR/

British Archaeology. http://www.britarch.ac.uk/ba/ba.html

British Journal of Sports Medicine. http://www.bjsportmed
.com/

British Medical Journal. http://www.bmj.com/current.shtml

Bryn Mawr Classical Review. http://ccat.sas.upenn.edu/
bmcr/

Cambridge Journals. http://titles.cambridge.org/journals/

Change. http://heldref.org/html/body_chg.html

Classical Antiquity. http://www.ucpress.edu/scan/ca-e/

Classical Review. http://www3.oup.co.uk/clrevj/

Clearing House. http://www.heldref.org/html/body_tch.html

College Teaching. http://www.heldref.org/html/body_ct.html

Comparative Literature and Culture. http://clcwebjournal.lib
.purdue.edu/

Configurations. http://muse.jhu.edu/journals/configurations/
index.html

Critique: Studies in Contemporary Fiction. http://www.heldref
.org/html/body_crit.html

Cultivate Interactive. http://www.cultivate-int.org/

Cultronix. http://eserver.org/cultronix/

Current. http://www.heldref.org/html/body_curr.html

Demokratizatsiya. http://www.heldref.org/html/body_dem
.html

Digital Humanities Quarterly. http://www.noraproject.org/
dhq/index.shtml

D-Lib Magazine. http://mirrored.ukoln.ac.uk/lis-journals/
dlib/

Early Modern Literary Studies. http://www.shu.ac.uk/emls/
emlshome.html

Education Policy Analysis Archives. http://seamonkey.ed.asu
.edu/epaa/

EJournal. http://www.ucalgary.ca/ejournal/

Electronic Journal of Analytic Philosophy. http://ejap.louisiana
.edu/

ELH. http://muse.jhu.edu/journals/elh/index.html

Esoterica: The Journal of Esoteric Studies. http://www
.esoteric.msu.edu/

European Review of History. http://www.tandf.co.uk/journals/
titles/13507486.html

European Review of Private Law. http://www.kluweronline
.com/issn/0928–9801/contents)
European Sociological Review. http://www3.oup.co.uk/
eursoj/
Explicator. http://www.heldref.org/html/body_expl.html
Explorations in Renaissance Culture. http://engphil.astate
.edu/ERIC/

Folklore. http://www.folklore.ee/folklore/
Forum Computerphilologie. http://computerphilologie.uni-
muenchen.de/
FQS—Forum: Qualitative Social Research. http://qualitative-
research.net/fqs/fqs-eng.htm

Germanic Review. http://www.heldref.org/html/body_ger
.html

Heldref Publications. http://www.heldref.org/
Helios. http://www.helios.org/
Heroic Age. http://members.aol.com/heroicage1/homepage
.html
HighWire Press. http://highwire.stanford.edu/
HINARI Health InterNetwork Access to Research Initiative.
http://www.who.int/hinari/en/
Historical Methods. http://www.heldref.org/html/body_hm
.html
History. http://www.heldref.org/html/body_his.html
Howard Hughes Medical Institute Bulletin. http://www.hhmi
.org/bulletin/

ImageText. http://www.english.ufl.edu/imagetext/
Indiana Journal of Global Legal Studies. http://ijgls.indiana
.edu/
Information Research. http://informationr.net/ir/
Information Society Journal. http://www.indiana.edu/~tisj/
International Health News. http://vvv.com/HealthNews/
International Journal on Digital Libraries. http://link.springer
.de/link/service/journals/00799/index.htm
Internet and Higher Education. http://www.elsevier.com/lo-
cate/iheduc)
Internet Archaeology. http://intarch.ac.uk/

JAC Online. http://nosferatu.cas.usf.edu/JAC/index.html

JAMA—Journal of the American Medical Association. http://jama.ama.-assn.org/

John Donne Journal. http://social.chass.ncsu.edu/jdj/

Journal of American Studies. http://www.baas.ac.uk/resources/jas/jas.asp)

Journal of Arts Management, Law, and Society. http://www.heldref.org/html/body_jamls.html

Journal of Business Administration Online. http://jbao.atu.edu/

Journal of Cell Biology. http://www.jcb.org/

Journal of Economic Education. http://www.heldref.org/html/body_jece.html

Journal of Education for Business. http://www.heldref.org/html/body_jeb.html

Journal of Educational Research. http://www.heldref.org/html/body_jer.html

Journal of Electronic Publishing. http://www.press.umich.edu/jep/

Journal of English and Germanic Philology. http://www.press.uillinois.edu/journals/jegp.html

Journal of Environmental Education. http://www.heldref.org/html/body_jee.html

Journal of Evolutionary Biology. http://link.springer.de/link/service/journals/00036/index.html

Journal of Experimental Education. http://www.heldref.org/html/body_jxe.html

Journal of Internet Banking and Commerce. http://www.arraydev.com/commerce/jibc/index.htm

Journal of Internet Technology and Secured Transactions. http://www.inderscience.com/browse/index.php?journalCODE=ijitst)

Journal of Military History. http://www.jstor.org/journals/08993718.html

Journal of Popular Film and Television. http://www.heldref.org/html/body_jpft.html

Journal of Seventeenth-Century Music. http://sscm-jscm.press.uiuc.edu/jscm/

Journal of Southern Religion. http://jsr.as.wvu.edu/

JSTOR. http://www.jstor.org/

Kairos: A Journal of Rhetoric, Technology, and Pedagogy.
http://english.ttu.edu/kairos/
Keats-Shelley Journal. http://www.luc.edu/publications/keats-shelley/ksjweb.htm
Kluwer Online. http://www.kluweronline.com/
Knowledge Quest: Journal of the American Association of School Librarians. http://www.ala.org/aasl/kqweb/index.html

Lion and the Unicorn. http://muse.jhu.edu/journals/lion_and_the_unicorn/index.html
Literary Imagination. http://www.bu.edu/literary/imagination/index.html
Logos. http://logosonline.home.igc.org/

Mäetagused. http://www.folklore.ee/tagused/
Mars Hill Review. http://www.marshillreview.com/
Mathematical Physics Electronic Journal. http://www.ma.utexas.edu/mpej/MPEJ.html
Medieval Forum. http://www.sfsu.edu/~medieval/
Medieval Review. http://www.hti.umich.edu/b/bmr/tmr.html
Medknow Publications. http:www.medknow.com/journals.asp)
Michigan Telecommunications and Technology Law Review. http://www. law.umich.edu/JournalsandOrgs/mttlr.htm
Mind. http://www.oup.co.uk/mind/
Minerva. http://www.ul.ie/~philos/
Mississippi Review. http://www.mississippireview.com/
MFS: Modern Fiction Studies. http://muse.jhu.edu/journals/mfs/index.html
Modernism/Modernity. http://muse.jhu.edu/journals/modernism-modernity/
MLN: Modern Language Notes. http://muse.jhu.edu/journals/mln.index.html
Montaigne Studies. http://humanities.uchicago.edu/orgs/montaigne/

naturalSCIENCE. http://naturalscience.com/ns/nshome.html
Nature. http://www.nature.com/
Negations. http://www.dataranglers.com/negations/index.html
New England Journal of Medicine. http://www.nejm.org/
Noûs. http://www.jstor.org/journals/00294624.html

Oak Ridge National Laboratory Review. http://www
.ornl.gov/info/ornlreview/
Oral History Review. http://www.ucpress.edu/journals/ohr/
Oxford American. http://oxfordamericanmag.com/
Oxford Art Journal. http://www3.oup.co.uk/oxartj/
Oxford University Press Journals. http://www3.oup.co.uk/
jnls/list/

Perspectives on Political Science. http://www.heldref.org/
html/body_pps.html
Philosophy and Literature. http://muse.jhu.edu/journals/
philosophy_and_literature/
Philosophy and Rhetoric. http://muse.jhu.edu/journals/par/
Poetics. http://www.elsevier.nl/inca/publications/store/5/0/
5/5/9/2/
Poetics Today. http://muse.jhu.edu/journals/poet/
Population and Development Review. http://www.jstor.org/
journals/00987921.html
Postmodern Culture. http://jefferson.village.virginia.edu/pmc/
contents.all.html
Project Muse. http://muse.jhu.edu/
Public-Access Computer Systems Review. http://info.lib.uh
.edu/pr/pacsrev.html

Renaissance Forum. http://www.hull.ac.uk/Hull/EL_Web/
renforum/
Renaissance Quarterly. http://www.jstor.org/journals/0034
4338.html
(Re)Soundings. http://www.millersv.edu/~resound/
Reviews in American History. http://muse.jhu.edu/journals/
reviews_in_american_history/index.html
Rhetorica: A Journal of the History of Rhetoric. http://www
.ucpress.edu/journals/rh/
Richmond Journal of Law and Technology. http://law.richmond
.edu/jolt/index.asp)
Richmond Review. http://www.richmondreview.com/
Romance Quarterly. http://ww.heldref.org/html/body_rq
.html

Science. http://www.sciencemagazine.org/
Science as Culture. http://human-nature.com/science-as-culture/

Scientist. http://www.the-scientist.com/
Sewanee Review. http://www.sewanee.edu/sreview.home.html
Sic et Non. http://www.cognito.de/sicetnon/
Society for Philosophy and Technology. http://scholar.lib
    .vt.edu/journals/SPT/spt.html
South Dakota Review. http://www.usd.edu/engl/SDR/
Speculum: A Journal of Medieval Studies. http://www.jstor.org/
    journals/00387134.html
Studies in Bibliography. http://etext.lib.virginia.edu/bsuva/sb/
Style. http://www.style.niu.edu/
Symposium. http://www.heldref.org/html/body_sym.html

Taylor & Francis Journals. http://www.tandf.co.uk/journals/
    alphalist.html
Teacher Talk. http://education.indiana.edu/cas/tt/tthmpg.html
Theatre Journal. http://muse.jhu.edu/journals/theatre_journal/
today@nasa.gov. http://www.hq.nasa.gov/office/pao/News
    Room/today/html

Vectors. http://www.vectorsjournal.org/
Victorian Literature and Culture. http://uk.cambridge.org/
    journals/vlc/
Victorian Poetry. http://muse.jhu.edu/journals/victorian_
    poetry/
Victorian Studies. http://muse.jhu.edu/journals/vic/

William and Mary Quarterly. http://index.umdl.umich.edu/
    journals/00435597.html
World Affairs. http://www.heldref.org/html/body_wa.html

URLs of Selected E-Zines

1099: The Magazine for Independent Professionals. http://www
    .1099.com/

Abalone Moon. http://www.abalonemoon.com/
Access to the Music Zone. http://www.amzmusiczine.com/
Agni Magazine. http://www.bu.edu/agni/
AlterNETive. http://www.ufsm.br/alternet/
American Prospect. http://www.prospect.org/
Apple Valley Review. http://www.applevalleyreview.com/

Areté. http://www.aretemagazine.com/
Art Bin Magazine. http://art-bin.com/aaehome.html
Astronomer. http://www.theastronomer.com/

Bathyspheric Review. http://www.montereybaypoetry.com/
Bellevue Literary Review. http://www.blreview.org/
Big Bridge. http://www.bigbridge.com/
Blackbird. http://www.blackbird.vcu.edu/
Blithe House Quarterly. http://www.blithe.com/
Blue Ear. http://blueear.com/index1.html

California Literary Review. http://www.calitreview.com/
Career Magazine. http://www.careermag.com/
Cashiers du Cinemart. http://www.impossiblefunky.com/
Cat World. http://www.catworld.co.uk/
Computer-Mediated Communication Magazine. http://www
    .december.com/cmc/mag/index.html
CyberKids. http://www.cyberkids.com/
Cyber Oasis. http://www.sunoasis.com/oasis.html

Dream People. http://www.dreampeople.com/
Drunken Boat. http://www.drunkenboat.com/

EContent Magazine. http://www.ecmag.net/
Edge. http://www.edge.org/
Elephant Rock Productions. http://www.erpmedia.net/
Exopa Terra. http://www.exopaterra.com/
Exquisite Corpse. http://corpse.org
Eyeshot. http://www.eyeshot.net/

fiction attic. http://www.fictionattic.com/
First Things. http://www.firstthings.com/
Flying Inkpot. http://inkpot.com/contents.html
Frauennews. http://www.frauennews.de/
From Now On. http://www.pacificrim.com/~mckenzie/
FrontPage Magazine. http://www.frontpagemag.com/
Full Moon. http://www.fullmoonlm.bravehost.com/

Global Inner Visions. http://www.give-zine.com/
Global Monitor. http://www.monitor.ca/monitor/
GoodBye!. http://www.goodbyemag.com/

Ink Magazine. http://www.ink-mag.com/
International Journal of Inclusive Democracy. http://www
.inclusivedemocracy.org/journal/

Keep On Truckin' Re-Visited. http://www.vipgrafx.com/hippy/
hippy.html
Kudzu. http://www.etext.org/Zines/Kudzu/

Like Water Burning. http://www.likewaterburning.com/
Literary Salt. http://www.literarysalt.com/
Little Magazine. http://www.albany.edu/~litmag/

MacUser. http://www.macuser.co.uk/
Mersey Beat. http://www.mersey-beat.net/
Mute Online. http://www.metamute.com/

nerve.com. http://www.nerve.com/
Newtopia: a magazine. http://www.newtopiamagazine.net/
Nova News Net. http://novanewsnet.ukings.ns.ca/
NovelAdvice Newsletter. http://www.noveladvice.com/

Onion. http://www.theonion.com/
Outland Ezine. http://members.tripod.com/jshultz/
Oyster Boy Review. http://www.oysterboyreview.com/

paperplates: a Magazine for 50 Readers. http://www.paper
plates.org/
Per Contra. http://www.percontra.net/
Pigdog Journal. http://www.pigdog.org/
Pig Iron Malt. http://www.pigironmalt.com/
Plexus. http://www.plexus.org/
pLUNGElit.com. http://www.plungelit.com/
Postmodern Village. http://www.postmodernvillage.com/
Prose Ax. http://ww.proseax.com/

Quest of the Unquietmind. http://unquietmind.com/
Quiet Mountain Essays. http://www.quietmountainessays.com/
RALPH. http://www.ralphmag.org/
Rose & Thorn. http://theroseandthorezine.com/

Scroll. http://www.scroll.org/
SeanAlonzo. http://seanalonzo.com/

sendecki. http://www.sendecki.com/
SPIKE Magazine. http://spikemagazine.com/index.html
Square Table. http://www.thesquaretable.com/
Stone Soup. http://www.stonesoup.com/
Story Bytes. http://www.storybytes.com/
Subverse. http://www.subverse.net/

Tarpaulin Sky. http://www.tarpaulinsky.com/
3AM Magazine. http://www.3ammagazine.com/
Toasted Cheese: The Online Literary Journal. http://www
    .toasted-cheese.com/ezine.htm
TREND Magazine. http://trendmag.com/

Underneath the Bunker. http://www.underneaththebunker.com/
Voices. http:www.1writersway.org/
Voltaire's Inkwell. http://www.voltairesinkwell.com/

Wild Violet. http://www.wildviolet.net/
Wine and Dine. http://www.winedine.co.uk/
Wolf Moon Press: A Maine Journal of Art and Opinion. http://
    www.wolfmoonpress.com/
Word. http://www.word.com/index.html
Wow Cool Comix. http://www.wowcool.com/

Websites

Websites of Companies
Offering Antivirus Software

Two kinds of virus-scanning software exist to protect computer
networks. One type is used on individual computers, while the
other scans the traffic going to and from the Internet at a gateway—looking for potentially harmful code. Both types of software not only try to remove any attached viruses, worms, or Trojans from software programs and documents but also inform
users about the actions being taken.

The software company Websites listed below offer a wide
range of antivirus and antiworm products for gateway-based and

local installations. These products differ primarily in their licensing and packaging schemes. It is important to note that these products are regularly tested by the computer security companies, and the companies' Website content is, therefore, continually updated to reflect the findings from those tests. Most of the listed companies also provide information services to their customers, as well as to the public, about newly released worms and viruses on the Internet. The Website content also details known countermeasures, commonly called patches or fixes.

Aladdin-Esafe. Alladin Knowledge Systems, Ltd. Headquarters: Arlington Heights, Illinois, and Tel Aviv, Israel. http://www .aladdin.com/

F-Secure. F-Secure, Inc. Headquarters: San Jose, California, and Helsinki, Finland. http://www.f-secure.com/

Network Associates—McAfee. McAfee, Inc. Headquarters: Santa Clara, California. http://www.mcafee.com/

Sophos. Sophos, Inc. Headquarters: Abingdon, Oxford, and Lynnfield, Massachusetts. http://www.sophos.com/

Symantec. Symantec Corporation. Headquarters: Cupertino, California. http://www.symantec.com

Trend Micro. Trend Micro, Inc. Headquarters: Cupertino, California. http://www.trendmicro.com

Websites of Companies Offering Firewalls

Firewalls are devices used to control the data traffic flowing into and out of a corporate network, to and from the Internet. By using firewalls, companies can block unwanted traffic and keep unauthorized users from gaining access to the network. Many of the available firewall products allow users to build a secure tunnel, or virtual private network, through the Internet so that two or more sites of an organization can use the Internet as a safe medium in which to communicate. Data are encrypted by one firewall before being sent, and the receiving firewall decrypts the data. This process minimizes the risks of various crack attacks. The firewall suppliers listed below have all provided well-designed products for these objectives in recent years.

Check Point. Check Point Software Technologies, Inc. Headquarters: Redwood City, California. http://www.checkpt.com/

Cisco. Cisco Systems, Inc. Headquarters: San Jose, California. http://www.cisco.com

NetScreen. Jupiter Networks, Inc. Headquarters: Sunnyvale, California. http://www.juniper.net/company/

Nortel. Nortel Networks. Headquarters: Brampton, Ontario. http://www.nortelnetworks.com/

SonicWall. SonicWall, Inc. Headquarters: Sunnyvale, California. http://www.sonicwall.com/

Watchguard. Watchguard Technologies, Inc. Headquarters: Seattle, Washington. http://www.watchguard.com/corporate/

Websites of Companies Offering Intrusion Detection Systems

Intrusion detection systems, abbreviated IDS, complete the set of technical precautions that organizations typically take to keep their computer networks safe. IDS assist system administrators in detecting whether a security breach has actually occurred in a computer network. IDS look for unwanted changes or data transfers in the network and note any anomalies. The following companies offer reputable intrusion detection systems.

Internet Security Systems. Internet Security Systems, Inc. Headquarters: Atlanta, Georgia. http://www.iss.net/

Tripware. Tripware, Inc. Headquarters: Portland, Oregon. http://www.tripwire.com

Websites Related to Sexual Assault, Cyber Pornography, and Cyber Child Pornography

Boston Globe. "Spotlight Investigation: Abuse in the Catholic Church." http://www.boston.com/globe/spotlight/abuse/.

This Website, powerfully disturbing to some, was created after the *Boston Globe* published a number of stories uncovering the scandals in the Boston area Catholic churches regarding sexual abuse of children. The site offers real-life documentary cases, message boards, and links to the problem in modern-day society.

CyberAngels
http://www.cyberangels.org/

This Website, composed of cyber volunteers, was the recipient of a 1998 Presidential Service Award. The CyberAngels was formed in 1995 and quickly became one of the most important cyber "neighborhood watch" groups safeguarding the Internet from stalkers, harassers, and child pornographers.

The CyberTipline
http://www.missingkids.com/cybertip/

This Website is designed to deal with leads from individuals wanting to report the sexual exploitation of children. Contact information is available to individuals wanting to assist in the fight against child sexual exploitation. Information provided is forwarded to law enforcement for investigation and review, and, when appropriate, to the Internet Service Provider.

Regulation of Child Pornography on the Internet
http://www.cyber-rights.org/reports/child.htm

This Website, compiled by Yaman Akdeniz, outlines the regulation of child pornography on the Internet, as well as cases and credible materials.

Torture Survivors Network
http://www.pacinfo.com/eugene/tsnet/

This Website, created by the nonprofit Rehabilitation Project for Survivors of Torture from Latin America and the Caribbean, today still functions as a forum for torture survivors, therapists, and the public.

WiredSafety Organization
http://www.wiredsafety.org/

This Website is the largest online safety, education, and assistance group in the world. They are considered to be an online "neighborhood watch" group, but they operate worldwide in cyberspace. In total, there are about 9,000 volunteers worldwide. They help online users deal with cyber-crime and online harassment, assist law enforcement around the globe to prevent and investigate cyber-crimes, educate online users about cyber-crime, and give information about various aspects of online safety, privacy, and security. With their affiliate, www.wired-cops.org, specially trained volunteers patrol the Internet, searching for child pornography, child molesters, and cyber-stalkers. Volunteers also search for and review family-friendly Websites, filter software products, and Internet services. The volunteers declare themselves a one-stop-shop for cyberspace safety.

Websites Related to Internet Periodicals and Journals

PubList.com
http://www.publist.com/about.html

PubList.com, the only Internet-based reference for more than 150,000 U.S. and international print and electronic publications, includes magazines, journals, e-journals, newsletters, and monographs. PubList.com provides a means for quick and easy access to detailed publication information, including titles, formats, publisher addresses, editor contacts, circulation data, and ISSN numbers. This Website also provides online users access to subscription services and article-level information through rights and permissions providers and document delivery services. The Website was founded in 1998 by Frederick Bowes, III, and in 2000 the heavily trafficked Website was acquired by Infotrieve to expand its service to information and research professionals and to assist in online ordering. In that regard, journal articles may be ordered through Infotrieve's online document delivery service, giving online users access to the complete text (formatted text and graphics) of most journal articles.

Internet-Related Audios, Documentaries, and Films

Audios

Chuck Harder Interviews Eric Jackson (2006)
http://www.archive.org/details/Chuck_Harder_Interviews_
Eric_Jackson.

This audio features Chuck Harder interviewing Eric Jackson, the writer of *The PayPal Wars.* Eric Jackson talks freely about the possibility of financial losses on the Internet and how thieves can use discarded canceled checks to create a PayPal account to steal money from checking accounts.

Cuba: We Want the Web! (February 9, 2003)
http://www.archive.org/details/Cuba_We_want_the_Web_1

In this 4-minute, 39-second audio, Thembi Mutch talks about his trip to Havana, Cuba, to find out not only why citizens there seem to have been gripped by a wave of enthusiasm for e-mail and the Internet but also why not everyone there seems to be able to get access to the Internet—and why people there are beginning to get suspicious. Does the Cuban government fear that citizens would become saturated by U.S. culture, or that the Internet will be used for subversive purposes? Or are the reasons more of an economic and a technical nature?

Department of Ongoing Digital Situations (April 8, 1994)
http://www.archive.org/details/DOODS_THE_SECRET_CITY

This audio is actually a mix that details some of the fascinating controversies in the early 1990s, when the Internet was in its early stages of development and when there were battles of competing interests between the online community and those in the corporate sector.

Governments Try to Control Internet (February 9, 2003)
http://www.archive.org/details/Governments_try_to_control
_Internet

Heather Ford, a leading expert on human rights and the Internet, speaks about how governments around the world censor the Internet because it is a powerful tool for dissenting voices.

In this 2-minute, 19-second audio, Ford outlines the latest government attitudes around the world regarding the Internet and the freedom of speech. She focuses on Burma, Zimbabwe, and Iran.

Interview of Tom Munnecke for Pew Internet Survey (September 17, 2005).
http://www.archive.org/details/UpliftAcademyInterviewof
 TomMunneckeforPewInternetSurvey.

In a production by Jenna Anderson of Elon University, Tom Munnecke is interviewed as part of the Pew-Elon University Internet Predictions Survey at the Accelerating Change Conference, held at Stanford University on September 17, 2005. More information on how 6 billion people can benefit from the Internet can be found at http://www.upliftacademy.org.

Jim Greenfield: Is the Government Watching You? (2006)
http://www.archive.org/details/Jim_Greenfield_Is_the_
 Government_Watching_You.

In this audio, Dave, a call-in guest from California, claims that he has government crackers blocking his telephone calls and following him to Internet blog sites. Jim, the show's host, thinks that Dave is likely a paranoid delusional. Would the government monitor a law-abiding citizen? Dave says "yes," because he blows the whistle on the government. This audio allows for a fascinating debate and lets the audience decide who is right—Jim or Dave.

Open Source in Educational IT (2006)
http://www.archive.org/details/WesleyAFryerOpenSourcein
 EducationalIT.

This podcast by Wesley A. Fryer features a skypecast interview with Afan Ottenheimer of JeoNet, an Internet applications development company based in Iowa City, Iowa. The interview focuses on the viability of using open-source software solutions in the educational Information Technology environment. Issues covered include security, total cost of ownership, support, and usability.

Podcasting for Fun and Profit (2006)
http://www.archive.org/details/Podcasting_for_Fun_and_Profit.

In this audio, Scott Paton discusses why podcasting can provide a huge opportunity for Website owners as well as ways to earn money with this new but largely untapped marketing medium. Another related Website can be found at http://coolwebtips .com/podcast.

The Doctor Will Google You Now with Thomas Lee, Susan Love and Deborah Wexler (2006)
http://www.archive.org/details/053105.

Produced by Open Source Media, this audio asks the interesting question, "Is the internet taking the mystery out of medicine?". Although access to health care has not improved vastly in some of the poorer regions around the globe, access to medical information online has improved. Thus, patients are now spending a lot of time online at PubMed to learn more about their health and health care issues. The effect of this Internet medium has been "sobering" to physicians who have enjoyed a centuries-old monopoly on medical terms and knowledge. This audio emphasizes that physicians, too, can gain from the Internet, for in the not-too-distant future, they will be diagnosing patients with the use of a wireless handheld like the BlackBerry.

Documentaries

Alle Kennis van de Wereld: Biography of Paul Otlet (Noorederlicht, 1998). This documentary, produced for Dutch television in 1998, features Paul Otlet, considered by many to be the father of information management. Narrated by W. Boyd Rayward, Otlet's biographer, the documentary focuses on the late 1800s and early 1900s when Otlet started what is today known as information science. Amazingly, one hundred years before the Internet came into being, Otlet used terms like "web of knowledge," "link," and "knowledge network" to describe his soon-to-be-developed vision for a central repository containing human knowledge.

BBS Documentary Interview Collection: Philip J. Kaplan (Bovine Ignition Systems, 2006). This documentary, directed and produced by Jason Scott Sadofsky, presents an interview with Philip J. Kaplan of FUCKEDCOMPANY.COM in 2002 for the BBS Documentary series. In New York City, Kaplan was asked questions about running the Internet forerunner, the Bulletin Board System, or BBS. Kaplan also spoke about pirate groups, modem costs, why a Web forum or BBS is useful for online users to post on, as well as tips for producing exciting conversations with online users. He also spoke about personality traits of BBS creators. Those wanting to view this documentary should know that profanity prevails. The director can be reached at Jason@textfiles.com.

BBS Documentary Interview Collection: Ted Beatie (Bovine Ignition Systems, 2006). This documentary, directed and produced by Jason Scott Sadofsky, presents a 2002 interview with Ted Beatie. Topics include using early modems, how students could fit into their schedules the dialing in of Bulletin Board Systems (BBSs), and a comparison of the Internet Relay Chat systems with the MIT Zephyr. The director can be reached at Jason@textfiles.com.

BBS Documentary Interview Collection: Minor Threat (Bovine Ignition Systems, 2005). This interview with Minor Threat was made in Austin, Texas, on January 30, 2004. The director and producer is Jason Scott Sadofsky. Although Minor Threat is known in the Internet community as the creator of Tone Loc, a war-dialer software program connecting thousands of phones looking for carriers and voices in a phone exchange, the documentary focuses on this individual's memories of public domain Bulletin Board Systems (BBSs) and cracking into them. The director can be reached at Jason@textfiles.com.

BBS Documentary Interview Collection: Tracer[ACiD] (Bovine Ignition Systems, 2005). Produced and directed by Jason Scott Sadofsky, this interview with Tracer, an artist for the ANSI art group known as ACiD, was made on March 8, 2004, in New York City. Interesting subjects covered include ANSI art groups, creating art in both ANSI and RIP computer graphics, and post-Bulletin Board System (BBS) events. The director can be reached at Jason@textfiles.com.

BBS Documentary Interview Collection: Mark Nasstrom (Bovine Ignition Systems, 2005). Produced and directed by Jason Scott Sadofsky, this interview with Mark Nasstrom was made in May 2003 in Seal Rock, Oregon. The documentary features discussion topics such as conducting election results using Bulletin Board Systems (BBSs), communications challenges in areas having sparse populations, and the *Boardwatch* magazine. The director can be reached at Jason@textfiles.com.

BBS Documentary Interview Collection: John Sheetz (Bovine Ignition Systems, 2005). Produced and directed by Jason Scott Sadofsy on March 22, 2003, in New Providence, New Jersey, this documentary focuses on teletype art and using Ham Radio to transfer text-based messages. The documentary covers such technical issues as creating RTTY art, the challenges of transferring artwork over Ham Radio, and some copyright issues. Sadly, the interviewee died before the documentary was released. The director can be reached at Jason@textfiles.com.

BBS Documentary Interview Collection: Loren Jones (Bovine Ignition Systems, 2005). Produced and directed by Jason Scott Sadofsky on January 24, 2004, this documentary moves away from the more technical side of Bulletin Board Systems (BBSs) and moves toward discussing the social and friendship issues surrounding BBSs. The interviewee was Loren Jones. The director can be reached at Jason@textfiles.com.

BBS Documentary Interview Collection: Ryan Brown (Bovine Ignition Systems, 2005). This interview, produced and directed by Jason Scott Sadofsky on May 22, 2003, in Mountain View, California, features Ryan Brown. Brown discusses how he ran a Bulletin Board System (BBS) as a very young man, how to get to know people strictly online, and how he won an award for running a BBS using creative software. The director can be reached at Jason@textfiles.com.

BBS Documentary Interview Collection: Rich Schinnell (Bovine Ignition Systems, 2005). Produced and directed by Jason Scott Sadofsy, this documentary features an interview with Rich Schinnell on February 29, 2004, in Rockville, Maryland. The discussion focuses on the 1982 time period, when the early IBM Personal Computer–compatible user groups were challenged by software and hardware issues. Rich Schinnell discussed his personal views on sharing software and profiting from public domain software. The director can be reached at Jason@textfiles.com.

Martial Law 9/11: Rise of The Police State (Jones Productions, 2005). Directed by Alex Jones and produced by Alex Jones and Kevin Booth, this controversial documentary discusses the rise of the police state in the United States since the terrorist attacks of September 11, 2001. Claims in the documentary are that the U.S.

Constitution has been shredded and that the United States now finds itself in a police state, run by a high-tech control grid set up across the United States. This documentary allegedly exposes who was behind the terrorist attacks, including the roots and history of its creators.

National Conference for Media Reform (Center for Media & Democracy, 2006). This 12-minute documentary, produced by the Center for Media & Democracy and distributed by MediaPolicyBlog.org, was designed to promote free speech, a diversity of voices, and an open and free Internet.

Recon 2005: Todd Macdermid on Encrypted P2P and VoIP Spaces with CUTLASS (www.recon.cx, 2005). Todd MacDermid is known in the Internet world for producing a number of open-source security tools, including steganographic network tunnels, encrypted mailing lists, and packet-mangling libraries. He has been a featured speaker at computer hacking conferences such as Black Hat in Las Vegas and Hackers on Planet Earth in New York City. In this documentary, MacDermid speaks about the lack of security surrounding Peer-to-Peer (P2P) file transfer, voice chat, and text-messaging systems. He also describes an open-source system known as CUTLASS, a tool powerful enough to support encrypted voice, chat, and file transfers.

Stop Identity Theft Now (Public Domain, 2004). This documentary features Sandy Klein, affiliated with the U.S. Attorney's Office in Los Angeles, California, who speaks on the fastest-growing crime on the Internet—identity theft. The documentary was produced to introduce the Stop Identity Theft Now Program and to answer the following questions: What is identity theft? How does an identity thief get personal information? How does an identity thief use personal information? How can citizens protect themselves from identity theft? What should victims do?

The Diary of a Network Administrator: Death of the Net (Leeware Development, 2005). In this documentary produced and directed by Lee Evans, the perceptions of a network administrator are presented. The latter discusses the Internet from its humble beginnings to its present-day state, where it is a primary medium for mass-marketing. The documentary also focuses on the problems created when there was increased access to the Internet by public citizens.

The Globalization of Time and Space 5 of 9 (University of Lethbridge Globalization Studies, 2004). This documentary, shown in lecture format, was directed by Anthony J. Hall and produced by Matt Sletto. The documentary focuses on a discussion of relationships between human beings and computers, including the logic used to make computers work. Other interesting social topics include how the Internet has made inroads into time and space and the challenges of expressing one's identity and sexuality in cyberspace.

The Internet and the Law (Ravensbourne College, 2004). Produced and directed by Ian Forrester, this documentary takes exception to the claim that the Internet is a sort of digital Wild West—an anarchic high-tech environment where there are few laws and regulations. The documentary argues that most of the laws applying to print and broadcast media apply equally to online media. Produced in seminar format, the documentary was made to raise awareness about some of the more important laws applying to online media.

Thomas Leighton: The Challenges of Delivering Content and Applications on the Internet (Northeastern University ACM, 2005). This documentary, directed by Ravi Sundaram and produced by Chris Lambert, features Thomas Leighton, the cofounder of Akamai Technologies in September 1998. Currently a chief scientist at Akamai Technologies, Dr. Leighton is considered to be a technological visionary and a top executive who determines the company's strategies. As one of the world's authorities on algorithms for network applications, Dr. Leighton's technological advances have earned him recognition as one of the Top 10 Technology Innovators in *U.S. News & World Report.* A professor at MIT, Dr. Leighton has served as the head of the Algorithms Group at MIT's Laboratory for Computer Science since its creation in 1996. As discussed in the documentary, Dr. Leighton holds many patents related to cryptography, digital rights management, and network algorithms. In 2004 he was elected into the National Academy of Engineering for contributions to the design of networks and circuits and for technology for Web content delivery.

Films

JuniorXDance Movie 2
Date: 2004
Length: 4 minutes, 28 seconds

Produced by the United Canadian Alliance clan and directed by JuniorX, this minidance movie was made in the World of Warcraft MMORPG, beta version. As with most movies produced in massively multiplayer games, the characters' actions are limited to e-motes and moves built into the game. In particular, JuniorX relies on Orc dance and fighting moves set to MC Hammer's tune "Can't Touch This."

Hackers
Date: 1995
Length: 107 minutes
Cast: Jonny Lee Miller, Angelia Jolie, Fisher Stevens, and
    Lorraine Bracco

This intriguing story focuses on a young cracker who accesses without authorization a highly secure computer system. Once in, he comes upon an embezzling scheme that is masked by a computer virus capable of destroying the world's ecosystem.

The Net
Date: 1995
Length: 114 minutes
Cast: Sandra Bullock, Jeremy Northam, Dennis Miller,
    Diane Baker

The story line revolves around a computer expert named Angela Bennett, a young, attractive computer analyst who is tied to her computer and her mother. An internet friend sent her a computer program to her to debug. The story gets interesting when the night he is to meet her, he is killed in an airplane crash. The story intensifies when Angela's computer files are erased and she is given a new identity.

Pirates of Silicon Valley
Date: 1999
Length: 95 minutes

Cast: Anthony Michael Hall, Noah Wyle, Joey Slotnick,
    John Di Maggio

The story line is about the creative ones in Silicon Valley: the creators of Apple Computers and Microsoft Corporation.

War Games
Date: 1983
Length: 114 minutes

This movie falls in the category of cyber-thriller. The story line is about a computer cracker who taps into the Defense Department's war computer. Lacking the intent to do so, he starts a confrontation of global proportions—World War III.

Warriors of the Net
Date: 2002
Length: 12 minutes, 40 seconds

This short animation film explains the way Internet Protocol packets flow on the Internet. The movie is available for information purposes by Ericsson Medialab. The film can be found at www.warriorsofthe.net/.

Television Series

Net Café, one of the world's most widely distributed television series, was broadcast in the United States as well as in more than 100 other countries from 1996 through 2002. Hosted by Stewart Cheifet, Jane Wither, and Andrew deVries, it covered the Internet revolution during the peak of the dot.com boom. The weekly series talked about the people and the culture behind the World Wide Web. The show gave Internet tips and helpful Websites (including Yahoo, Google, and eBay), talked about Internet startups, and featured interviews with the pioneers behind the development of the Internet. The winner of a number of journalistic excellence and broadcast awards, the show was produced on location at various internet cafes around the Silicon Valley and the San Francisco Bay areas.

Glossary

Access Controls The physical or logical safeguards preventing unauthorized access to information resources.

Actus Reus One of the four elements on which Anglo-American law bases criminal liability. This element means a criminal action or a failure to act when one is under a duty to do so.

Algorithm A set of rules and procedures for resolving a mathematical or logical problem, much as a recipe in a cookbook helps baffled cooks in the kitchen resolve meal problems. A computer program can be viewed as an elaborate algorithm, and in computer science, an algorithm usually indicates a mathematical procedure for solving a recurrent problem. Information security professionals are concerned with cryptographic algorithms—those used to encrypt, or encode, messages. Different algorithms have different levels of complexity, which is related to key size. For example, a 41-bit key is twice as hard to crack, or decode, as a 40-bit key. A 128-bit key is a trillion times harder to crack than a 40-bit key.

Anonymous Remailers A computer service that privatizes e-mail and typically contains the sender's identity. Anonymous remailers send e-mail messages that arrive in a receiver's inbox without a sender's identity.

Antivirus Software Detects viruses and notifies the user that the virus is present. This type of software keeps a database of "fingerprints," a set of characteristic bytes from known viruses, on file.

ARPAnet The first transcontinental, high-speed computer network, built by the U.S. Department of Defense as an experiment in digital communications.

Attendant Circumstances One of the four elements on which Anglo-American law bases criminal liability. This element means the existence of certain necessary conditions. With some crimes, it must be proven that certain events occurred, or certain facts are true, for a person to be found guilty of the crime.

285

Black Hat Hackers Individuals who engage in destructive computer exploits, motivated by revenge, sabotage, blackmail, or greed. These exploits can and often do result in harm to property or persons.

Blog Short for Web log, it is an online journal and forum for commentary that doubles as a public discussion board. Blogs are Web pages on which an individual posts diarylike entries. Blogs are gaining popularity, particularly as a means of marketing one's talents online—a replacement for old-fashioned paper resumes.

"C" The computer language created in the 1970s by Dennis Ritchie.

Code The portion of the computer program that can be read, written, and modified by humans.

Council of Europe Convention on Cybercrime The first global legislative attempt of its kind to set standards on the definition of cyber-crime and to develop policies and procedures governing international cooperation to combat Internet crime.

Cracking Gaining unauthorized access to computer systems to commit a crime.

CyberAngels A not-for-profit organization of White Hat hackers assisting targets of cyber-crimes, particularly of cyber-stalking and cyber-pornography. This organization also helps legal authorities to fight Internet crime.

Cyber-Ethics By definition, ethics applied to the online environment. Although cyber-ethics has become an important topic for elementary school children, high school students, college and university students, and for those in the workplace in recent years, the treatment of what is and is not cyber-ethical behavior varies from place to place. The general principle is "do unto others online as you would have them do unto you."

Cyber-pornography Using the Internet to possess, create, import, display, publish, or distribute pornography (especially child pornography) or other obscene materials.

Cyberspace Composed of hundreds of thousand of interconnected computers, servers, routers, switches, and fiber optic cables, cyberspace allows the critical infrastructures—telecommunications, energy, banking, water systems, government operations, and emergency services—to work. Thus, cyberspace is the "nervous system" of the global economy.

Cyber-stalking Using cyberspace and the Internet to control, harass, or terrorize a target to the point that he or she fears harm or death to self or others close to the target.

Cyber-terrorism Unlawful attacks and threats of attack by terrorists against computers, networks, and the information stored within when done to scare or coerce a government or its citizens to further the perpetrator's political objectives.

Cyber-vigilantism Using the Internet to conduct vigilante activities.

Decryption or Decipher This is the process of taking encrypted data that has been put into a "secret" format called ciphertext and converting it to the original, readable plaintext. To complete this process, a key or password is needed.

Diffie-Hellman Public-Key Algorithm (DH) An algorithm, the DH was developed by Whitfield Diffie and Martin Hellman in 1976. The DH, an algorithm upon which a number of secure connectivity Protocols on the Internet are built, is now celebrating more than thirty years of use. In short, DH is a means of securely transmitting a secret to be shared between two parties over an untrusted network in real time. A shared secret is critical for two parties who likely have not communicated before, so that they are able to encrypt communications. Today, DH is used by protocols such as Internet Protocol Security (IPSec), Secure Shell (SSH), and Secure Sockets Layer (SSL).

Directory Search Engines These search engines do not search on the Internet for information but obtain it from individuals who enter it into the search engine's databases. Because each directory search engine has its own way of categorizing information, multitudes of them exist.

Distributed Denial of Service (DDoS) A cyber-attack whereby a cracker bombards a targeted computer with thousands or more fake requests for information, causing the computer to run out of memory and other resources. Therefore, it is either slowed down dramatically, or stopped altogether. In this exploit, the cracker uses many—often hundreds or thousands—of previously cracked computers connected through the Internet to originate the attack. The multiple origins of the cyber-attack make it especially difficult to defend against.

Domain Name System (DNS) The Domain Name System, or DNS, is a hierarchical system of naming hosts and placing these hosts into categories. The DNS is a way of translating numerical Internet addresses (which are hard to recall) into word strings (which are easier to recall) to denote user names and locations.

E-mail Mail sent from one party to another through an electronic system.

Encryption The mathematical conversion of data into a form from which the original information cannot be restored without using a special key. In other words, algorithms are used.

Ethernet In 1985, the U.S. Institute of Electrical and Electronic Engineers (IEEE) developed standards for Local Area Networks (LANs) called the IEEE 802 standards. These standards presently form the basis of most networks. One of the IEEE 802 standards—the IEEE 802.3—is known as "Ethernet," the most prevalently used LAN technology around the globe. Ethernet was designed by the Xerox Corporation in 1972, and in its simplest form, it used a passive bus operated at 10 Mbps. A 50-ohm coaxial cable connected the computers in the network.

File Transfer Protocol (FTP) A protocol used to transfer files between systems over a network.

Firewalls Programs used to provide additional security for networks by blocking access from the public network to certain services in the private network.

Fixes or Patches Updated system software, created by computer security experts to close security gaps discovered in software after it has been released into the marketplace.

Flooding A type of Internet vandalism resulting in Denial of Service to authorized, legitimate users of a Website or computer network.

Hacker According to those in the Computer Underground (CU), a hacker is a person who enjoys learning the details of computer systems and how to stretch their capabilities to do creative things. Media professionals often use the word incorrectly to signify a cracker.

Hacktivism Using the Internet to promote one's own political platform or mission.

HTML (Hyper Text Markup Language) A Web browser interprets HTML, the programming language used to code Web pages on the Internet, into words and graphics so that users can view the pages in their intended layout and rendering.

HTTP (Hyper Text Transfer Protocol) A programming language used to transfer data over the World Wide Web. All Internet Website addresses begin with http://. When a computer user types a URL, or Uniformed Resource Locator, into the browser (a software application used to locate and display Web pages) and hits the "Enter" key, the computer sends an HTTP request to the correct Web server. The Web server, designed to handle these kinds of requests, then sends the user the desired HTML (Hyper Text Markup Language) page.

Identity Theft Stealing and then misusing someone else's identity (often, using their social security number or credit card numbers) to commit crimes. Identity theft may or may not employ computers.

Information Packet A piece of data of fixed or variable size sent through a communications network like the Internet. A message sent over a network is broken into smaller data packets that are disassembled and then reassembled.

Infringing Intellectual Property Rights and Copyright Infringing Intellectual Property Rights (IPR) and Copyright can occur online and, thus, falls in the broad-based category of "cyberspace theft." An example would be copying another person's work from an online source without being authorized to do so or without giving proper credit for the work—including songs, articles, movies, or software.

Intellectual Property (IP) IP treats the protection of creative products of the human mind like the protection of property. IP laws, in particular, grant certain kinds of exclusive rights over these creative products on the analogy of property rights.

Internet A network connecting computer systems.

Internet Protocol (IP) The reason that networked computers can communicate with one another is a common communication protocol called Internet Protocol, or IP. All applications utilized on the Internet have been designed to make use of this IP—in earlier times and in the present.

Internet Service Providers (ISPs) ISPs are companies that, for a fee, provide online users with a software package, a username, a password, and an access phone number so that, equipped with a modem, they can log onto the Internet, browse the World Wide Web, or send and receive e-mail. Lately, ISPs have been offering high-speed services.

Internet Usage Policies Before giving employees access to the company's Internet service, companies require would-be Internet users to sign an "Internet Usage Policy" form, which has them agree that they will be accountable for their online activities.

Internic An organization called the Internic maintains a database having all registered domains for the world. Therefore, anyone can query its database by means of a whois. Although several organizations maintain whois databases, the Internic has the main database. Thus, any company, institution, or organization wanting to have its own domain name must register it with Internic.

Intranet or Intranet Site An information system internal to an organization that is built with Web-based technology.

Intrusion Detection System (IDS) An Intrusion Detection System, or IDS, is a security service that determines how computer system cracks or attempted cracks are detected in real-time or in near-real-time and provides warnings. Intrusion detection analysts review logs and other available network information to look for suspected or real intrusions. IDSs are commercially available.

Intrusion Recovery Recovering from a computer network incident, whereby a cracker or a set of crackers has exploited vulnerabilities in a computer network, usually to cause harm to the network.

IP Address A numerical identifier divided into a part identifying a network (such as a company, an educational institution, or a government agency) and another part identifying each computer in the network.

Thus an IP address is much like a street name and a house number in a nonvirtual neighborhood.

Linux An operating system used prevalently on Internet servers and utilized by many large corporations as an alternative to the Microsoft operating system.

Listserv An online subscription to some topic of interest.

Local Area Network (LAN) Networks often contained in one or more buildings in nearby locations.

Malicious Code Software programs like viruses and worms designed to exploit weaknesses in computer software. They may replicate and attach themselves to other programs.

Mens Rea One of the four elements on which Anglo-American law bases criminal liability. This element means a culpable mental state.

MMORPG (or Massively Multiplayer Online Role-Playing Game) A form of computer entertainment played by one or more individuals using the Internet. It is commonly known as online gaming.

MOO An acronym for MUD, Object-Oriented.

Moore's Law Since the 1960s, the number of transistors per unit area has been doubling every one-and-a-half years—thus increasing computer power. This progression of circuit fabrication is called Moore's Law, named after Gordon Moore, a pioneer in the integrated circuit field and creator of the Intel Corporation.

MUD A multiuse dungeon scenario employed in computer gaming.

Newsgroups Users connected to the Internet can get articles posted to thousands of discussion groups called newsgroups. The articles are topic-arranged and are distributed through an electronic Bulletin Board System called Usenet.

Phreaking Crackers using technology to their advantage to get free long-distance telephone calls.

Piracy Illegally copying or distributing software using the Internet.

Podcasting Technology making it easy for Internet users to create their own audio recordings and then post them on the World Wide Web.

Privacy The state of being free from unauthorized access.

Privacy Laws Laws that deal with the right of individual privacy, critical to maintaining the quality of life that citizens in a free society expect. Privacy laws generally maintain that an individual's privacy shall not be violated unless the company (or the government) can show some compelling reason to do so—such as by providing evidence that the safety of the company (or of the nation) is at risk.

Prohibited Result One of the four elements on which Anglo-American law bases criminal liability. This element means a result that brings harm to property or to persons.

Protocol A set of rules governing how communications between two programs has to take place to be considered valid. Because information is sent over the Internet in information packets that are disassembled and then reassembled, the Transmission Control Protocol (TCP) and the Internet Protocol (IP) are critical for the proper transmission and routing of these information packets.

Remote Retrieval A common use of the Internet today is for users to search for and retrieve information located on remote computers. There are primarily three ways that this search and retrieval on the Internet can be accomplished, one of which is using the File Transfer Protocol (or FTP) to transfer files between systems over the network, particularly from a host (that is, server) to a remote computer (that is, client). Commercial browsers like Netscape, for example, have built-in FTP capabilities.

Robot A robot uses a software program to search, catalog, and then organize information on the Internet.

Root Servers A group of thirteen servers worldwide that are responsible for the basic level of the domain name system.

Routers Specialized computer devices at the border of an Internet-connected network that store a specialized map of the Internet and contribute to this map by telling its neighbors about what it knows about its part of the Internet.

Security Being protected from one's adversaries, particularly from those who would do one harm—intentionally, or otherwise, to property or to a person. Information Technology security issues, in particular, include but are not limited to authentication, critical infrastructure protection, disaster recovery, intrusion detection and network management, malicious code software protection, physical security of networks, security policies, the sharing of rights and directories, and wireless security.

Server A computer program carrying out some task on behalf of a user—such as delivering a Website page or delivering e-mail. Computers on which these server applications are run are also called servers.

Social Engineering A deceptive process, whereby computer crackers "engineer" a social situation to trick others into giving them access to an otherwise closed network.

Spammers Individuals who send unsolicited e-mails to targets over the Internet for commercial purposes. There is typically a criminal intent to defraud the targets or to get them interested in some Website they would typically avoid—such as those dealing with child pornography.

Spoofing A type of appropriation of an online user's identity by others online, causing fraud or attempted fraud in some cases, as well as critical infrastructure breakdowns in other cases.

Telnet A terminal emulation program, or a program based on that protocol, allowing online users to log onto the Internet. That is, Telnet is an Internet application allowing a user's Personal Computer (PC) to act as a terminal to a remote system. Telnet requires that the user be familiar with UNIX operating system software.

Trademark Law Trademark law governs disputes between business owners over the names, logos, and other means they use to identify their products and services in the marketplace. It is interesting to note that more than 33,000,000 Internet domain names have been registered, including tens of thousands of domain names apparently infringing trademark and service marks.

Trade Secret Law Trade secret laws protect trade secrets. In the United States, for example, a trade secret can be a number of things—devices, formulas, ideas, and processes—that give the owner a distinct market advantage. Trade secrets could be movie scripts, customer lists, or special types of computer hardware. For that reason, the owner would want to have some trade secret protection to ensure that the public or competitors could not get the trade secrets by improperly accessing files containing those secrets (that is, proprietary information) stored in computers.

Trojan (general term) Given its moniker after the Trojan Horse of ancient Greek history, a Trojan is a particular kind of network software application developed to stay hidden on the computer where it has been installed. Like worms, Trojans generally serve malicious purposes and are in the "malware" classification. Trojans sometimes access personal information stored on home or business computers and then send it to a remote party via the Internet. Alternatively, Trojans may serve merely as a "backdoor" application. Trojans can also launch DoS attacks. A combination of firewalls and antivirus software should be used to protect networks against Trojans.

Trust A complex concept studied by scholars from a number of academic disciplines, defined to be present in a business relationship when one partner willingly depends on an exchanging partner in whom one has confidence. The term *depend* can take on a number of meanings in this context, including the willingness of one partner to be vulnerable to the actions of the other, or the expectation of one partner to receive ethically bound behaviors from the other partner. Security issues regarding Information Technology center on maintaining trust in e-commerce transactions.

Trusted Operating System or Secure Operating System Trusted and secure operating systems are significant terms in the information security profession. The basis of this terminology is that clients can place their trust in the people and in the organization operating a trusted system. Technically, a trusted, or secure operating system refers to one labeled as "hardened OS" (hardened operating system) or "trusted OS" (trusted operating system). Although the primary objective of both of these is to provide a secure operating environment, each takes a different approach for meeting that objective.

UNIX A widely used computer operating system, UNIX has a standardized and well-publicized set of rules and interfaces governing the interactions of humans and programs. Therefore, it is considered to be an "open" operating system rather than a proprietary system (such as a Microsoft product), in which these rules and interfaces are not so easily accessible.

Virus An often harmful computer program that is capable of replicating itself by embedding a copy of itself in other computer programs. A virus is often sent through attachments in e-mail.

VoIP (Voice over Internet Protocol) Technology allowing telephone calls to be placed over networks like the Internet—making it less expensive than conventional land lines to place long-distance phone calls.

W3C Founded by Tim Berners-Lee, the W3C is a consortium of industry leaders wanting to promote standards for the World Wide Web's (that is, the Internet's) continued development and for greater interoperability between Internet products.

White Hats Computer hackers having good intentions; they tend to hack into systems with authorization to find flaws in the computer network that could be exploited by unwanted cyber-intruders (called crackers or the Black Hats).

Whois (general term) Whois is a TCP/IP utility allowing system administrators to query compatible servers to get detailed information about other Internet users.

Wide Area Network (WAN) A network like the Internet that connects physically distant locations.

Wiphishing An act executed when an individual covertly sets up a wireless-enabled laptop computer or access point to get other wireless-enabled laptop computers to associate with it before launching a crack attack.

Wired Equivalency Protocol (WEP, or Wi-Fi technology standard) WEP, a protocol adding security to wireless LANs based on the IEEE 802.11 Wi-Fi standard, is an OSI Data Link layer technology that can be turned "off" and "on." WEP was developed to give wireless networks the same level of privacy protection as a wired network. WEP is formulated on a security notion called RC4, using a combination of system-generated values and secret user keys. The first implementations of

WEP supported only 40-bit encryption and had a key length of 40 bits and 24 additional bits of system-generated data, resulting in a 64-bit total. Since its inception, computer scientists have determined that 40-bit WEP encryption is too weak—which is why product vendors today utilize 128-bit encryption (having key lengths of 104 bits) or higher. Wireless network devices utilize WEP keys to encrypt the data stream for communications over the wire.

Worms Self-replicating computer programs that are self-contained and do not need to be part of another program to replicate. In contrast, a virus attaches itself to and becomes part of another executable program.

XML (or Extensible Markup Language) An index developed by Tim Bray that makes online auctions possible. XML allows computer programmers to attach "tags" or universal codes to distinguish, say, business names from telephone numbers.

Zombie A computer program awaiting a signal from a cracker to bombard a particular Website. On command, multiple zombies can simultaneously transmit thousands or more of fake requests for information to the targeted Website. As the computer tries to handle these requests, it soon runs out of memory. It either slows down dramatically or stops altogether.

Index

About the Author

Bernadette H. Schell is dean of the Faculty of Business and Information Technology at Ontario's only laptop university, the University of Ontario Institute of Technology in Oshawa, Ontario, Canada. Dr. Schell is the 2000 recipient of Laurentian University's Research Excellence Award. Formerly the director of the School of Business and Administration at Laurentian University, Dr. Schell authored four books with Quorum Books on organizational and personal stress, corporate leader stress and emotional dysfunction, stalking, and computer hacker profiling. She has also coauthored *Cybercrime: A Reference Handbook*, part of ABC-CLIO's Contemporary World Issues series. Dr. Schell's *Webster's New World Hacker Dictionary*, published by Wiley, was released in September 2006. Dr. Schell also serves on a number of boards, including the Greater Toronto Marketing Alliance.